Logic-based Nonlinear Image Processing

Also Available from SPIE Press Exclusively Published by New Age International Pvt. Ltd.

Fourier-Transform Spectroscopy Instrumentation Engineering
Vidi Saptari
ISBN: 978-81-224-2823-0, Original Edition Price: $ 48.00, Indian Edition Price: Rs. 395.00

Tissue Optics : Light Scattering Methods and *Instruments for Medical Diagnosis,* 2nd Ed.
Valery Tuchin
ISBN: 978-81-224-2829-2, Original Edition Price: $ 99.00, Indian Edition Price: Rs. 995.00

Local Approximation Techniques in Signal and Image Processing
Vladimir Katkovnik, Karen Egiazarian, Jaakko Astola
ISBN: 978-81-224-2830-8, Original Edition Price: $ 99.00, Indian Edition Price: Rs. 695.00

Logic-based Nonlinear Image Processing
Stephen Marshall
ISBN: 978-81-224-2825-4, Original Edition Price: $ 51.00, Indian Edition Price: Rs. 395.00

Introduction to the Optical Transfer Function
Charles S. Williams, Orville A. Becklund
ISBN: 978-81-224-2831-5, Original Edition Price: $ 78.00, Indian Edition Price: Rs. 695.00

Digital and Analog Fiber Optic Communications for CATV and FTTx Applications
Avigdor Brillant
ISBN: 978-81-224-2832-2, Original Edition Price: $ 115.00, Indian Edition Price: Rs. 999.00

Infrared Fibers and Their Applications
James A. Harrington
ISBN: 978-81-224-2827-8, Original Edition Price: $ 78.00, Indian Edition Price: Rs. 495.00

Optoelectronics of Solar Cells
Greg P. Smestad
ISBN: 978-81-224-2826-1, Original Edition Price: $ 46.00, Indian Edition Price: Rs. 395.00

MOEMS: Micro-Opto-Electro-Mechanical Systems
Manouchehr E. Motamedi
ISBN: 978-81-224-2833-9, Original Edition Price: $ 113.00, Indian Edition Price: Rs. 795.00

Introduction to Image Stabilization
Scott W. Teare, Sergio R. Restaino
ISBN: 978-81-224-2822-3, Original Edition Price: $ 51.00, Indian Edition Price: Rs. 395.00

Fundamentals of Polarimetric Remote Sensing
John R. Schott
ISBN: 978-81-224-2834-6, Original Edition Price: $ 68.00, Indian Edition Price: Rs. 395.00

Fundamentals of Infrared Detector Materials
ichael A. Kinch
N: 978-81-224-2824-7, Original Edition Price: $ 61.00, Indian Edition Price: Rs. 395.00

Logic-based Nonlinear Image Processing

Stephen Marshall

Tutorial Texts in Optical Engineering
Volume TT72

SPIE PRESS
Bellingham, Washington USA

NEW AGE

First Indian Sub-Continent Edition : 2010
by New Age International (P) Ltd., Publishers

Branches:

- 36, Malikarjuna Temple Street, Opp. ICWA, Basavanagudi, **Bangalore**. ✆ (080) 26677815
- 26, Damodaran Street, T. Nagar, **Chennai**. ✆ (044) 24353401
- CC-39/1016, Carrier Station Road, Ernakulam South, **Cochin.** ✆ (0484) 4051303
- Hemsen Complex, Mohd. Shah Road, Paltan Bazar, Near Starline Hotel, **Guwahati**. ✆ (0361) 2543669
- No. 105, 1st Floor, Madhiray Kaveri Tower, 3-2-19, Azam Jahi Road, Nimboliadda, **Hyderabad**. ✆ (040) 24652456
- RDB Chambers (Formerly Lotus Cinema) 106A,1st Floor, S.N. Banerjee Road, **Kolkata**. ✆ (033) 22275247
- 16-A, Jopling Road, **Lucknow**. ✆ (0522) 2209578
- 142C, Victor House, Ground Floor, N.M. Joshi Marg, Lower Parel, **Mumbai**. ✆ (022) 24927869
- 22, Golden House, Daryaganj, **New Delhi**. ✆ (011) 23262370, 23262368

Printed in India at V. K. Printers, Delhi.
C-09-11-4197

ISBN: 978-81-224-2825-4
Original Edition Price: $ 51.00
Indian Edition Price: Rs. 395.00

PUBLISHING FOR ONE WORLD
NEW AGE INTERNATIONAL (P) LIMITED, PUBLISHERS
4835/24, Ansari Road, Daryaganj, New Delhi-110002
Visit us at **www.newagepublishers.com**

Introduction to the Series

Since its conception in 1989, the Tutorial Texts series has grown to more than 70 titles covering many diverse fields of science and engineering. When the series was started, the goal of the series was to provide a way to make the material presented in SPIE short courses available to those who could not attend, and to provide a reference text for those who could. Many of the texts in this series are generated from notes that were presented during these short courses. But as stand-alone documents, short course notes do not generally serve the student or reader well. Short course notes typically are developed on the assumption that supporting material will be presented verbally to complement the notes, which are generally written in summary form to highlight key technical topics and therefore are not intended as stand-alone documents. Additionally, the figures, tables, and other graphically formatted information accompanying the notes require the further explanation given during the instructor's lecture. Thus, by adding the appropriate detail presented during the lecture, the course material can be read and used independently in a tutorial fashion.

What separates the books in this series from other technical monographs and textbooks is the way in which the material is presented. To keep in line with the tutorial nature of the series, many of the topics presented in these texts are followed by detailed examples that further explain the concepts presented. Many pictures and illustrations are included with each text and, where appropriate, tabular reference data are also included.

The topics within the series have grown from the initial areas of geometrical optics, optical detectors, and image processing to include the emerging fields of nanotechnology, biomedical optics, and micromachining. When a proposal for a text is received, each proposal is evaluated to determine the relevance of the proposed topic. This initial reviewing process has been very helpful to authors in identifying, early in the writing process, the need for additional material or other changes in approach that would serve to strengthen the text. Once a manuscript is completed, it is peer reviewed to ensure that chapters communicate accurately the essential ingredients of the processes and technologies under discussion.

It is my goal to maintain the style and quality of books in the series, and to further expand the topic areas to include new emerging fields as they become of interest to our reading audience.

Arthur R. Weeks, Jr.
University of Central Florida

This book is dedicated to my late parents:

To my father, William George Marshall, and to my mother, Clara Marshall, for their kindness and encouragement right up to the end of their lives.

This book is dedicated to my late parents:

To my father: William George Marshall, and to my mother: Clara Marshall, for their kindness and encouragement right up to the end of their lives.

Contents

Acknowledgments

There are many people who have convinced me of the need for this book and have encouraged me to write it. However, the principal motivator of this work has been Professor Ed Dougherty of Texas A&M University. He has gently goaded me into action by his belief that there is a gulf between the mainly mathematical texts describing morphology and the engineers and computer scientists who implement solutions. He convinced me that I could write a book that would go some way towards bridging this gulf.

I must also thank all of my research students past and present—in particular, Neal Harvey, Mahmoud Hamed, Neil Woolfries, Alan Green, and Kenneth Hough, examples of whose work have been included in this book. I also thank Lyndsay Fletcher of the University of Glasgow for her input to Chapter 9 as well as my other research students: Druti Shah, George Matsopoulos, Peter Kraft, Bjorn Rudberg, Jennifer McKenzie, Santiago Esteban Zorita, and Wei Yi.

I would also like to thank all of the people I have worked with over the years, including Hans Burkhardt, Moncef Gabbouj, Ioannis Pitas, Murat Kunt, Jean Serra, Fernand Meyer, Etienne Decenciere Ferrandiere, Giovanni Sicuranza, Gianni Ramponi, Ed Coyle, Gonzalo Arce, and Lou Scharf. And I cannot forget my colleagues John Soraghan and Tariq Durrani.

I also thank Timothy Lamkins and Beth Huetter at SPIE for their assistance in making this book happen.

Lastly, I would like to thank my wife Joan for her patience.

Stephen Marshall
October 2006

Chapter 1
Introduction

Classical signal and image processing uses linear processing techniques. These are methods based in the familiar Fourier, Z, and Laplace transforms. These methods assume that signal and image data may be processed by mapping them onto lower-dimensional orthogonal spaces resulting in solutions designed by decomposing the input into sinusoidal components and processing them individually. While mathematically elegant, this imposition of linearity results in a very limited set of processing operations compared to the total set of solutions possible, i.e., both linear and nonlinear. For example, techniques based on rank ordering of values, logical and geometric processing approaches can give excellent results, particularly for image processing applications. This approach should not be viewed as an alternative to the classical methods, but as a superset of techniques containing many new novel techniques as well as the linear techniques listed above.

The model chosen to convey these concepts is that of digital logic. This is because it can quite literally capture any processing operation, linear or nonlinear, that may be required. Many engineers and computer scientists are comfortable with its notation and concepts. Minimization techniques and software tools are available to reduce complex solutions into their simplest form, and the solutions translate readily into electronic hardware or software implementations.

Every digital signal or image processing operation can be viewed at its most basic level as the manipulation of a series of finite-length binary strings. Whether the operation is implemented on a processor through software or in dedicated hardware, the data and the algorithms are invariably mapped through electronic logic components, which are inherently binary in nature.

Therefore, every digital signal and image processing task can be cast in terms of a logical representation. It does not matter if the data is binary, grayscale, color, or multiband, nor whether the operation is linear or nonlinear. If it can be programmed, then it can be placed in the context of a logical representation.

In nonlinear image and signal processing, the design of operators is carried out by seeking the optimum mapping from one set of binary strings to another. This contrasts with the linear approach which formulates a solution by optimizing coefficients within a generalized multiply-accumulate context. It should be noted, however, that even this linear method is then mapped into digital logic for computation.

In these terms, linear models may be perceived as restricted subsets within a logical framework. Hence, a nonlinear solution to the same problem will be a more general result that will be either better or the same as the linear solution, provided that other conditions are met. One of the most important conditions is that sufficient training data is available.

So why do linear solutions remain so common? There are a number of reasons. The first is familiarity. Engineers and signal processors are trained in linear techniques and are reluctant to depart from the security of these familiar solutions unless the subsequent improvements are great.

Also, the superposition properties of linear models makes parameter estimation straightforward. This means that a small number of examples of system behavior may be used to infer performance across a range of conditions. In theory, a linear system may be completely described by observing the same number of training examples as the rank of the system. In practice, even allowing for the system observations to be noisy, the model may be fully characterized with only a small amount of over-determination. Also, if the linear system model is extended by adding extra parameters, only a linear increase in the number of training examples is required.

The situation is much more complex for nonlinear systems. The task is to seek the optimal logical mapping from all possible mappings. No simple superposition properties exist, and in the most general unconstrained design case, every combination of input variables must be observed a sufficient number of times in order to estimate the conditional probabilities of the output. Extending the system model by adding more parameters leads to a rapid increase in the size of the required training set. This contrasts sharply with the linear problem where it is only required that one estimate the autocorrelation matrix, which is a much smaller set of values than the conditional probabilities.

For logical mappings containing a large number of variables, the required training set may be impossibly large. It may well be that even after observing a huge set of training examples, some combinations have not been observed or have been observed an insufficient number of times to make a statistically accurate estimate of their conditional probabilities.

In the face of these estimation difficulties, it is not surprising that linear methods remain popular. Also, in many problems such as circuit analysis and audio applications, linear solutions are quite satisfactory. These systems are inherently linear with their steady state and transient behavior being completely modeled as a product of sinusoids and decaying exponentials. Other systems make much use of Gaussian noise models and these sit naturally in a linear context. In these cases there is no need to look any further, this model is satisfactory.

However, these linear approaches that work so well for many problems are not necessarily as useful for image processing applications. The 2D nature of image processing problems combined with human visual perception often requires more involved decisions than is the case in 1D signal processing. For example, the tasks might include object and texture classification or size distribution estimation. In many cases, the 2D image is a single projection of the 3D world via unspecified models with unknown parameters. The additional problems of perspective, shadow, and occlusion lead to further ambiguities that can only be resolved with the application of experiential knowledge.

Visual perception is a complex task; it is not tolerant of the linear approximations that arise from frequency decomposition and the projection of signals into orthogonal subspaces. As a result of the perceptual importance of edges, the essential components of images tend to occupy a wide range of the frequency domain. The corrupting noise processes may well overlap the signal in such a way as to make linear separation impossible. It is also difficult to quantify image quality through simple measures such as mean-absolute error (MAE) and mean-square error (MSE).

For example, an image may be restored in such a way that it contains only a tiny variation in MAE from some ideal original, but if the higher frequency components are lost or there is significant phase distortion, it may look very poor to a human observer. On the other hand, large variations in brightness and contrast (leading to large error measures) may be tolerable provided that the edges are distinct.

Despite these points, linear image processing techniques have thrived because of their mathematical elegance and their ability to describe continuous signals. Also, the process of sampling such that continuous signals are represented only by their values at discrete points may be completely described by linear mathematics.

Despite this, there are strong arguments for seeking solutions to image processing problems in terms of logical mappings. Consider a linear "image-to-image" processing task, which might include restoration, noise reduction, enhancement, or shape recognition.

We begin with a signal that is sampled in three dimensions (two spatial and one intensity). Let us assume that the image is $256 \times 256 \times 8$ bits. Whatever processing is to be carried out, the result will eventually be mapped back into the same discrete signal space. The bits within the finite strings of the input image are interpreted as part of an unsigned binary number in order to be given an arithmetic meaning. In most linear operations, such as filtering, the unsigned integers will be converted to real or complex numbers containing a mantissa and an exponent. In order to compute the various linear multiply-accumulate transformations, these numbers are then mapped into electronic circuits and viewed as finite-length binary strings. The circuits operate at their most basic level by employing digital electronics to carry out Boolean algebra on the binary strings to produce different binary strings.

The resulting binary values are then mapped back to real or complex numbers that are eventually clipped and quantized into the $256 \times 256 \times 8$ bit signal space that forms the output image.

So even though we may have carried out a fundamentally linear operation such as a Fourier or wavelet transform, it has been implemented as a series of logical operations. We have mapped the signal in terms of binary strings through digital logic to a resulting set of binary strings. However, we have in effect imposed *linearity constraints* such that at every stage of processing the following two statements are true:

1. The binary strings being manipulated have a direct interpretation in terms of real or complex numbers.
2. The logical operations applied to the strings are restricted to those that carry out equivalent linear operations, such as multiplication and addition of real or complex numbers.

Nonlinear image processing is presented here as a generalization of the above operation by removing the linearity constraints. It seeks the optimum mapping implemented directly in logic. The linear solution should be viewed as a special case of the set of all logic-based solutions rather than as an alternative. Given this generalization, the optimum nonlinear solution will be either better or equivalent to the linear solution, but it should not be worse. This inequality holds regardless of the problem or the criteria, provided that the training data is sufficient.

The above argument has lead to various researchers in this field issuing the provocative claim that "all image processing is nonlinear."[1]

The principal reason for adopting this strategy is to see if the other solutions available through a logical approach are useful and offer advantages over linear solutions. Linear solutions can be easy to compute. It is not difficult to derive the optimum linear smoothing filter for an image with noise, but the result of applying this filter is an image which is invariably blurred, causing a loss of signal information. Here, a nonlinear solution such as the median filter gives much better results leading to noise removal and edge preservation without blurring, despite the fact that the median filter takes no account of the image or noise statistics.

In removing the linear constraint, the process of finding the optimum solution becomes much more difficult to compute. However, if the consequences of linear processing are unacceptable results, we must try to do this.

The work in this area has focused on the design of filters. Many applications are possible within this context such as noise reduction, shape, character and object recognition, enhancement, restoration, texture classification, spatial and intensity sampling, and rate conversion.

In practice, all filters are limited in some way. These limits are known as constraints. For example, the filter designed for a particular application may be constrained to lie in a particular class. The optimum filter is therefore the best filter within that class. In this work, we seek the optimum filter from the class of filters that have a logical implementation. This also includes morphological and rank-order filters (which may be cast in the above context and therefore may provide solutions that have an interpretation in terms of shape or numerical ordering).

Linear filters require little training data. In theory, only the same number of examples as the number of parameters is necessary to determine a solution. However, for nonlinear filters, the training process amounts to the estimation of the conditional output probabilities. In the most general case, each training example only provides information about one specific combination of input variables. It is not possible to infer anything about the behavior of the filter for other sets of inputs. For a stochastic system, a sufficient number of observations of every input combination would be required to arrive at a robust estimate. The number of input combinations grows rapidly as the number of input variables increases.

In order to be able to design the filter from a realistically sized training set, further constraints must be applied to the filter. The filter is an estimator; it uses the input values to estimate an unobserved quantity. By making simple assumptions about the image statistics, we can estimate the output value at a specific point by considering only a finite window of observations centered at that point. For binary values, the output becomes a logical function of the input variables. If the window contains n points, there are 2^n combinations of input variables for which the relevant output must be estimated. Therefore, there are 2^{2^n} possible functions (or filters) and it is the objective of the design process to determine which one of these functions corresponds to the optimum.

Among the 2^{2^n} functions that may be applied within an n point window, there will be many subclasses of functions. We may decide to restrict the choice to a filter that is idempotent or increasing. *Idempotence* implies that the filter has only a one-off effect on the image such that repeated application of the filter leads to no further modification of the image. *Increasing* implies that the filter preserves signal ordering. It can be shown that *increasing* filters map to logical functions that contain no complementation of the input variables. This drastically reduces the size of the training set required and therefore makes filter design easier. This can be explained in terms of logic (since a much smaller set of functions is under consideration) or in terms of statistical estimation (since now a single training example may be used to infer information about other combinations of input variables).

If we assume that the statistics of the image are wide-sense stationary, then we may assume that the same optimum function applies at every point in the image. The filter then becomes translation-invariant. This not only simplifies the processing, but in effect increases the available training data because we do not distinguish between data collected at different locations in the image.

Nonlinear filters can be effective in retaining structural information while removing background clutter in a way not possible with linear operations. They can often be application-specific.

Historically, nonlinear filters have developed along three independent strands: morphological, rank-order, and stack. However, all can be brought together and expressed in the context of logic.

Mathematical morphology has its roots in shape.[2,3] A signal is probed by a structuring element to determine if it "fits" inside the signal. Mathematically, it has been expressed in set theory as explained by Minkowski. Initially, the work grew from binary images, although it can equally well be applied to 1D signals and has since been extended to grayscale[4] and complete lattices.[5]

Morphology was developed in the context of set theory. It does, however, take little more than a change in notation to show that the basic operation of erosion corresponds directly to a logical AND of the input variables. For all practical purposes, what is called an *erosion* in morphology is called a Minkowski subtraction in set theory. It is also called an *intersection* in mathematics and in digital electronics it is called an AND function. Matheron made the observation that every increasing, translation-invariant set operator may be represented as a union of erosions. To an electronics engineer this means that all operators can be implemented as a sum of products (and they do not require complementation). The building blocks of mathematical morphology such as erosion, dilation, opening, closing, and their repetitions under unions and intersections all have straightforward implementations in digital logic.

The second historical line came from the field of rank-order-based filters. These are inherently grayscale in nature and have at their core the ordering of the variables within an input window into their rank order. Trivial examples are the maximum and minimum but the success story of these filters is the median. It possesses powerful noise-removal properties and requires no knowledge of signal and noise distributions. It can be shown to be the optimum estimator of samples in unbiased noise for an MAE criteria.

The final strand of nonlinear filtering is stack filters which are based on Boolean logic operations applied within a finite window. They process grayscale signals by thresholding them at a number of levels and filtering the resultant stack of binary signals with a logic function.

The three types of filters have the following relationship:

$$\text{Order-statistic filters} \subset \text{stack filters} \subset \text{morphological filters}$$

In other words, morphological filters are the most general of the three, stack filters are a subset within morphological filters, and order-statistic filters are a subset within stack filters.

The literature describing the above methods tends to be quite academic and mathematical. It is the purpose of this book to bring these methods together and explain them in terms of logical operations. The objective is to bring these techniques to a whole new community, namely electronic engineers and computer scientists. The text assumes a basic knowledge of logic minimization such as could be achieved through simple K maps. It also uses very basic statistics to identify the optimum filters in the examples given.

The remainder of the book is structured as follows:

Chapter 2 introduces the concept of logic-based image processing through a document restoration example. Chapter 3 considers methods of evaluating the errors in filtering and gives more examples of document processing including resolution changing, edge noise, and optical character recognition. Chapter 4 looks at filter training and the trade-off between the different types of errors. Chapter 5 develops the relationship between logic-based image processing and mathematical morphology and introduces increasing filters. Chapter 6 establishes the link between logic-based image processing and certain classes of order-statistic filters involving variations on the median. Chapter 7 extends these concepts to grayscale through the model of computational morphology. Chapter 8 describes how each of the classes of filters may be implemented in electronic hardware. Chapter 9 presents a case study on image processing of astronomical images. Lastly, Chapter 10 presents conclusions.

With this new perspective on image processing, let us consider a number of applications starting with document restoration.

References

1 E. R. Dougherty and J. Astola, *An Introduction to Nonlinear Image Processing*, SPIE Press, Bellingham, WA (1994).

2 G. Matheron, *Random Sets and Integral Geometry*, Wiley, New York (1975).

3 J. Serra, *Image Analysis and Mathematical Morphology*, Academic Press, New York (1982).

4 J. Serra, *Image Analysis and Mathematical Morphology*, vol. 2, Academic Press, New York (1988).

5 H. J. Heijmans, *Morphological Operators*, Academic Press, New York (1994).

Chapter 2

What Is a Logic-Based Filter?

When a fuzzy fax or faded photocopy is received, it is usually possible to figure out what it says although the text may be badly damaged or corrupted by noise. This is because the human brain has knowledge of the characters and fonts of the alphabet and is able to fill in the gaps (or ignore noise) using experience. Over time, the brain has learned roughly what to expect and can correct it. The image in Fig. 2.1 is a text document that has been corrupted with 10% additive noise. This type of corruption is called salt-and-pepper noise.

It is clear that the document could be typed out again to reproduce the original version precisely. Therefore it is possible to restore the document fully, using intelligent human intervention. It might be more difficult to do this for unfamiliar alphabets such as Chinese or Arabic if the person had no previous knowledge of the shapes of the characters.

The important question is this: Is it possible for a computer program to "learn" this process? The answer is "Yes," certainly to a very large extent.

Consider the following approaches:

- Employment of a standard filter such as the median or positive/negative median;
- Heuristic approaches for estimating a good filter;
- The use of statistics to identify the optimal filter out of all filters.

Many image processing specialists will immediately reach for the median filter on seeing the type of noise present in Fig. 2.1.[1] The median filter takes the pixels within a small window, places them in rank order and selects the middle one. In this text document, the image has only two levels, 0 and 1 (or black and white). The median in this case corresponds to a simple count of the pixels in the window. If more than half are black, then the output is set to black, otherwise it is set to white. There are two main disadvantages to this approach. The first is that the median is a *dual operator*. This means that its effect on black pixels is exactly mirrored by its effect on white pixels. In this case, we have only additive noise and so the ideal filter

Figure 2.1 Corrupted text document. This document contains 10% additive salt-and-pepper noise.

should remove only black pixels to restore the image to its original state. Unfortunately, the median treats both equally, so it cannot simply remove black pixels while leaving white unchanged. The second disadvantage is that the median filter carries out exactly the same operation for all images and all noise distributions. Therefore, it cannot possibly be the best filter in all these cases. There must be better filters possible, and it would be reasonable to assume that these will be different for images with different structure and corrupting noise.

This leads to the second approach above: heuristic methods. This is basically designing by guesswork—a human designer tries out different filters to improve the quality of the resulting image. A typical approach might be to observe that the noise in the image consists of isolated black pixels and hypothesize that if these pixels could be identified and switched to white, most of the noise would be removed.

This would constitute a rule such as: "Switch a black pixel to white if it has more than *N* white neighbors," where *N* could be anything from 1 to 8. Another rule might include structural details. For example, "Switch every black pixel to white that has a white pixel immediately above and below it." These rules may give some improvement especially in simple cases. However, they may be difficult to formulate in more complex images especially in areas where the noise and signal detail are very similar. It would also be impossible to know if the filter obtained was the best one out of all available filters or whether it might be improved.

The third approach is to use statistics to determine the *optimum* filter. Consider the following very simple pair of images shown in Fig. 2.2 with the original image I_o on the left and the noise corrupted version I_n on the right.

The ideal image consists of a number of horizontal bars. The noisy image has been corrupted by a random noise process that has both added pixels to the back-

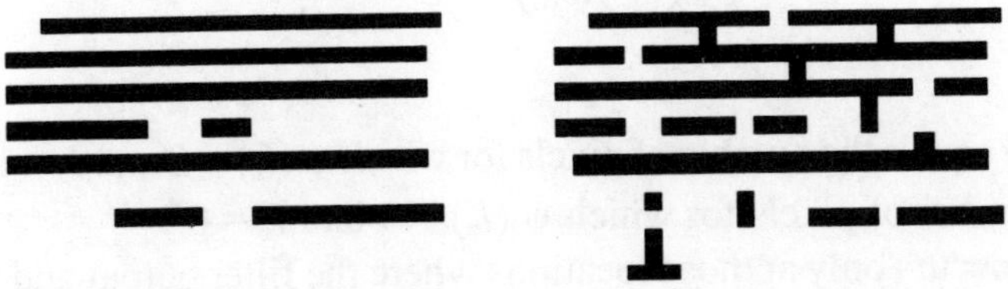

Figure 2.2 A simple example with the original image I_o on the left and the noise corrupted version I_n on the right. The images differ by 26 pixels.

ground and also subtracted pixels from the foreground. (It is assumed here that the black pixels are the foreground and white are the background). The total number of pixels differing between the two images is 26.

Starting with the noise-corrupted image I_n, the objective is to find a filter ψ to recover the original image I_o. In practice, this may not be possible. The design task therefore reduces to finding the optimum filter ψ_{opt} out of all possible filters ψ that minimizes the difference between the filtered noisy image $\psi(I_n)$ and the original I_o. In the language of statistics, an optimal estimator is being sought. Its task is to estimate the true value of the image pixels from a noise-corrupted version.

2.1 Error Criterion

Thus far, the words *best* and *optimum* have been used loosely and have not been given any specific mathematical definition. In quantitative terms, they require a measure of similarity. The measure usually used in this context is the mean-absolute error (MAE). Another measure is the mean-square error (MSE), and the relative merits can be debated for grayscale images. For binary images, MAE and MSE are identical.

Given two images $I_1(r, c)$ and $I_2(r, c)$ with the same number of R rows and C columns, their MAE is defined as

$$MAE(I_1, I_2) = \frac{1}{RC}\sum_{c=0}^{C-1}\sum_{r=0}^{R-1}\left|I_1(r,c) - I_2(r,c)\right|. \tag{2.1}$$

The optimum filter is therefore defined as the one that minimizes the difference between the ideal image I_o and the filtered version of the noisy image $\psi(I_n)$,

$$MAE(\psi(I_n), I_o). \tag{2.2}$$

For binary images, the MAE consists of just two types of errors:

$$MAE\left(\psi(I_n), I_o\right) = \frac{\Delta(0,1) + \Delta(1,0)}{RC}, \tag{2.3}$$

where $\Delta(0, 1)$ equals the number of pixels for which $\psi(I_n) = 0$ and $I_o = 1$, and $\Delta(1, 0)$ equals the number of pixels for which $\psi(I_n) = 1$ and $I_o = 0$.

An error occurs only at those locations where the filter output and the ideal image differ. For each location where this occurs, the contribution to the total MAE is precisely one pixel.

Note: The MAE gives equal weighting to $\Delta(0, 1)$ errors and $\Delta(1, 0)$ errors (i.e., pixels that should have been set to black but have been missed, and those that have been set to black but should not have been). There may be cases where different weightings for these two types of errors may be appropriate.

2.2 Filter Constraints

Any practical filter that can be designed to operate on an image must be constrained in some way. There are an infinite number of possible filters that may take many different forms. An unconstrained filter for image restoration would be absurdly large. Consider the image shown in Fig. 2.3. An unconstrained filter would require that every output pixel had a different filter, and require that every one of those filters was a function of every pixel in the input image. This would be a true optimal filter. In fact, it would be a number of different filters, since each pixel would require its own estimator. Such a totally unconstrained filter would clearly be impractical. However, it is possible to constrain the problem to make it practical and at the same time produce acceptable results even though the filter used would be suboptimal.

Two commonly used constraints are the *window constraint* and *translation invariance*.

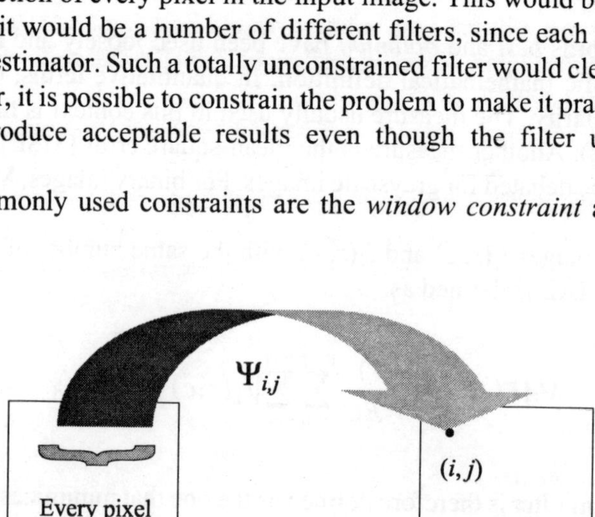

Figure 2.3 Unconstrained Filter. An unconstrained filter requires that every output pixel has a different filter and every one of these filters is a function of every pixel in the input image. This type of filter is unrealistic and practical results are usually achieved by using filters that are both windowed and translation invariant.

2.3 Window Constraint

An assumption is made that the noise and signal statistics of the processed image are *localized*. In simple terms, a pixel is more likely to be related to its immediate neighbors than to pixels a large distance away. This means that it is not necessary to consider every location of the input image when estimating the value of a pixel. The filter is therefore influenced mainly by *local* structure.

The true value of a pixel may therefore be estimated from the noise-corrupted version of the image by considering only a finite collection of pixels within a local neighborhood centered on the pixel. Considering pixels outside of the neighborhood will add little further information. If this assumption is not true or only partially true, the filter obtained will be suboptimal. If the size of the window is increased, the resulting filter will be closer to the optimal. The images in this book will therefore be processed using a sliding window (or mask) of values centered on the pixel to be estimated.

2.4 Translation Invariance

An assumption is usually made that the statistics of the image detail and the corrupting noise process are wide-sense stationary. This means that the same filter may be used at every location of the image. If this assumption is not true, the filter produced will be a weighted average of the different filters that would be optimum at each location. In this case the solution would be suboptimal. In practice, the results obtained from filters based on this assumption have been acceptable for a wide variety of applications.

Therefore, adopting these two constraints means that not only will the images be processed using a sliding window, but the filter characteristics within the window will be the same for all locations in the image.

Note: It is true that a window-based filter is unable to determine pixel values at the edges and corners of the image. For image-restoration purposes, the edge pixels are usually simply omitted from the process. In applications such as image coding where processed pixels are required, a smaller asymmetrical version of the window is used at the extreme locations.[2]

2.5 Filter Windows

A number of different windows of increasing size have been commonly used in image restoration. Some examples are shown in Fig. 2.4. For the same data, the *best* possible filter for a large window will always be better (or the same) than the *best* filter that may be found within a smaller sub-window.

Therefore, if $MAE\ (\psi^{i}_{\text{opt}})$ is the mean-absolute error that results from filtering an image with the optimum filter using a window containing i pixels, then

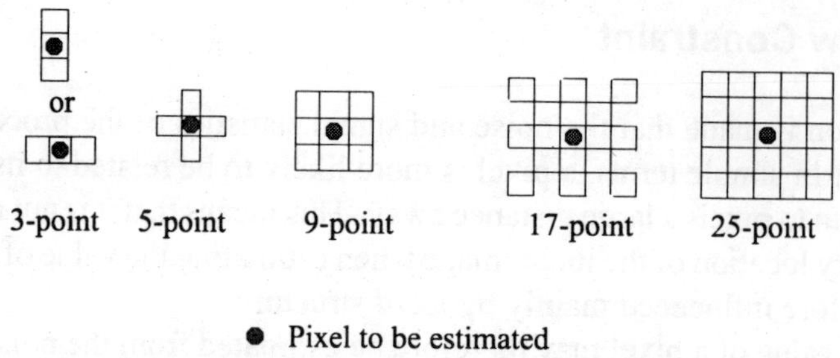

Figure 2.4 Examples of filter windows. Filter windows usually have an odd number of pixels so that the pixel to be estimated may be at the center.

$$MAE\,(\psi^{i}_{\text{opt}}) \leq MAE\,(\psi^{j}_{\text{opt}}) \quad \text{for } i > j \tag{2.4}$$

It would appear that the best strategy is therefore to use the largest possible window to give the minimum MAE and hence the best restoration. In theory this is true. However, in practice the optimal filter may be difficult to determine for a large window, which can mean that on balance it is better to use a smaller mask. This will be discussed in more detail in Chapter 4.

2.6 Filter Design

The task of filter design is to determine the optimum filter within a sliding window, e.g., for the problem described in Fig. 2.2. In this illustrative example, the simple three-point horizontal window shown in Fig. 2.5 will be used.

The pixel values in the window may be considered as an input vector of binary values $\mathbf{x} = (X_0, X_1, X_2)$. The filter output is an estimate of the value at the center of the window (location corresponding to X_1). The filter output value may therefore be represented as a binary function of the three input pixel values.

There are eight ($= 2^3$) possible combinations of the input vector $\mathbf{x}$. The design process of the filter ψ consists of allocating a value of either 0 or 1 for each possible combination of $\mathbf{x}$. There are 256 ($= 2^{2^3}$) different ways of doing this, and therefore 256 different filters.

The optimum filter ψ_{opt_i} is the one yielding the lowest value of MAE of all of these possible filters. It would be possible to filter the noisy image with every one of the 256 filters and to compare them to the ideal image and calculate the MAE.

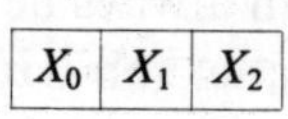

Figure 2.5 Three-point horizontal filter window.

X_2	X_1	X_0	N_0=0	N_1=1
0	0	0	64344	3
0	0	1	18	1
0	1	0	10	1
0	1	1	2	11
1	0	0	16	3
1	0	1	1	4
1	1	0	1	12
1	1	1	2	87

From I_n From I_o

Figure 2.6 Design strategy. A table of observations is constructed by sliding the filter window through the noisy image I_n and for each location, either column N_0 or N_1 is incremented depending on whether the value of the corresponding pixel in the ideal image I_o is 0 or 1, respectively.

However, even if the designer was willing to do this, the process does not scale well and is impractical for anything other than very small filter windows.

Fortunately, there is a more intelligent strategy for identifying the optimum filter (see Fig. 2.6). The key feature here is the table of observations. The three columns on the left show all combinations of the input variables $\mathbf{x} = (X_0, X_1, X_2)$.

This table is constructed by sliding the three-point window through image I_n. All of the values in the two columns on the right are initially set to zero. At each location, the pixel values within the input window correspond to one line in the table. If the corresponding pixel value in the ideal image I_o is 0, then the value in column N_0 is incremented; if it is 1 then the value in column N_1 is incremented. This is repeated for every location in the image. At the end of this process the two columns on the right, N_0 and N_1, indicate the number of times that the value of the corresponding pixel observed in the ideal image I_o was 0 or 1 for each input combination $\mathbf{x}$.

At the end of the process, the two columns on the right contain counts of the number of times the corresponding pixel in the ideal image was either 0 or 1 for each input combination.

The resulting table can be used to:

- Design the optimum filter,
- Measure its error, and
- Measure the increase in error by using any suboptimal filter.

2.7 Minimizing the MAE

Consider the line in the table of observation corresponding to $\mathbf{x} = (1, 0, 1)$. As previously explained, a pixel value of 1 corresponds to black and 0 corresponds to white.

(a)

X_2	X_1	X_0	N_o	N_1
.	.	.	.	.
1	0	1	1	4
.	.	.	.	.

(b)

Figure 2.7 (a) Window content for $\mathbf{x} = (1, 0, 1)$ and (b) count of observations for input $\mathbf{x} = (1, 0, 1)$.

The content of the filter window is as shown in Fig. 2.7(a). The count of observation values for this window is repeated in Fig. 2.7(b). As the filter window is passed over the image, the pixel pattern shown in Fig. 2.7(a) was observed a total of five times. It can be seen that on four occasions the corresponding ideal value was 1, and in the remaining case it was 0. For this type of filter, a single value of output must be assigned to each input combination. If the filter output for this particular input is set to 1, it will be correct for four pixels and cause an error at just one. Alternatively, setting the filter output to 0 would cause four pixels to be in error and only one to be correct.

Hence, the allocation of the output value for each input combination (i.e. the design strategy) is as follows:

$$\psi_{\text{opt}}(\mathbf{x}) = 1 \quad \text{if} \quad N_1(\mathbf{x}) \geq N_0(\mathbf{x}) \quad \text{and} \quad \psi_{\text{opt}}(\mathbf{x}) = 0 \text{ otherwise,} \tag{2.5}$$

where $N_1(\mathbf{x})$ and $N_0(\mathbf{x})$ correspond to the number of observations in the ideal image I_o for which the corresponding pixel was 1 or 0, respectively.

In other words, the output is set to the value that is correct most often. This process is repeated for every input combination, and hence the optimum filter may be determined. Figure 2.8 shows the table for all inputs. The output of the optimum filter corresponds to the most commonly occurring ideal value.

Using only the most basic knowledge of Boolean algebra, the filtering function can be easily shown to be

$$\psi_{\text{opt}}(\mathbf{x}) = X_0X_1 + X_0X_2 + X_1X_2. \tag{2.6}$$

Num	X_2	X_1	X_0	N_o	N_1	OPTIMUM F
0	0	0	0	64344	3	0
1	0	0	1	18	1	0
2	0	1	0	10	1	0
3	0	1	1	2	11	1
4	1	0	0	16	3	0
5	1	0	1	1	4	1
6	1	1	0	1	12	1
7	1	1	1	2	87	1

Figure 2.8 Optimum filter output for observations shown.

It is in fact the majority function that requires at least two of the input pixels to be 1 for the output to be 1. In the binary case, this also corresponds to the median filter. This is the best filter that can be obtained within the three-point window and no other function can give a smaller MAE for this window.

The optimally restored image is shown in Fig. 2.9 and has 14 pixels in error compared to the ideal image I_o. Recall that the noisy image I_n had 26 pixels in error.

The optimally restored image is still not completely fixed. In fact, a reduction in error from 26 pixels to 14 might seem significant, but is far from perfect. It should be noted that the restoration window is very simple and tests have shown

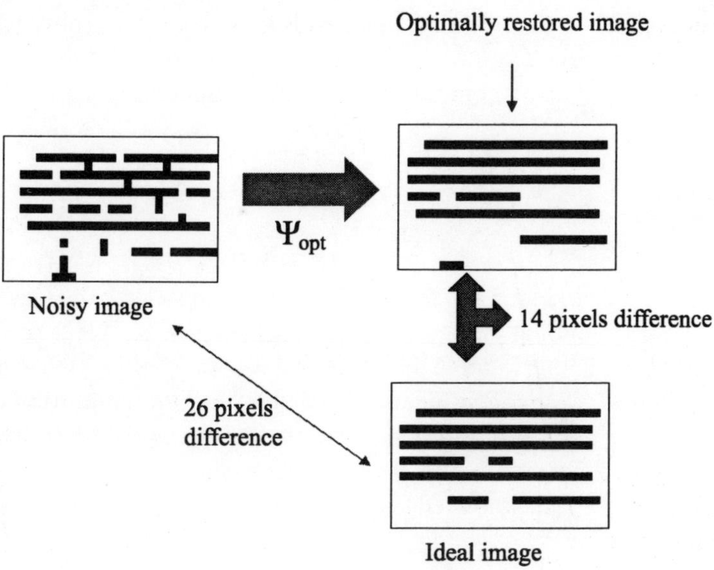

Figure 2.9 The optimally restored image (with a three-point window) is shown at the top right. It has 14 pixels in error compared with 26 in the noisy image.

that for a larger window, such as a 5 × 5, the image may be restored to within one pixel of the original.

The error of 14 pixels could have been calculated by comparing the restored image $\psi_{opt}(I_n)$ to the ideal image I_o and counting the pixels that differ. However, this and other measures may be obtained from the table of observations, which is the topic of the next chapter.

2.8 Summary

This chapter has shown an example of a simple restoration process using an optimum filter. It has defined what is meant by constraint and error criterion. It has shown how a table of observations may be constructed from noisy and ideal test images, and it has shown how the logical function defining the optimum filter may be determined from the table of observations. This table may also be used to determine various other errors and to compare different types of filters, the subject of the next chapter.

References

1 D. R. K. Brownrigg, "The weighted median filter," *Commun. ACM*, **27**, 807–818 (1984).

2 M. Ghanbari, *Video Coding: An Introduction to Standard Codecs*, IEE London (1999).

Chapter 3

How Accurate Is the Logic-Based Filter?

3.1 Optimum Filter Error

In most practical situations, even the optimal filter within a finite window is unable to recover the original image exactly. However, the remaining error will be the smallest possible for that window. Returning to the observations for the simple example shown in Fig. 2.8, it can be seen that for the line of the table discussed, **x** = (1, 0, 1), the output of the filter was set to 1. While this gave the correct output value for four of the pixels, it still left one in error. In general, each input **x** will make a contribution to the error equivalent to the smaller of the two values of N_0 or N_1. The error from each input may be totaled to give the overall filter error.

The MAE arising from the optimum filter is therefore

$$MAE(\psi(I_n), I_o) = \frac{\sum \min(N_o, N_1)}{RC} \tag{3.1}$$

for an image containing R rows and C columns.

Num	X_2	X_1	X_0	N_o	N_1	OPTIMUM F	Error (pixels)
0	0	0	0	64344	3	0	3
1	0	0	1	18	1	0	1
2	0	1	0	10	1	0	1
3	0	1	1	2	11	1	2
4	1	0	0	16	3	0	3
5	1	0	1	1	4	1	1
6	1	1	0	1	12	1	1
7	1	1	1	2	87	1	2
						Total	14

Figure 3.1 Calculation of error after filtering with the optimal filter. The shaded areas represent the error arising from each input.

Figure 3.2 An error image. The figure shows the pixels not repaired following filtering with an optimal filter.

For the simple example shown in Chapter 2, the error calculation is shown in Fig. 3.1. The smaller value in the two columns, either N_0 or N_1, is added to the total and it can be seen that the overall number of pixels in error is 14. Figure 3.2 shows the error image, the correctly restored pixels, and shows the errors marked on the image. It can be seen that all isolated error pixels have been correctly restored. However, where a number of adjacent pixels are in error, the filter has been unable to correct them, which is not surprising as it only operates within a three-point window.

When a suboptimal filter function is used to filter the image, the MAE increases relative to the optimal. The amount by which the error increases may be computed from the table of observations. The error only increases for those inputs where the suboptimal filter has a different output to the optimal. For these inputs, it increases by the difference between N_0 and N_1. For example, when the noisy image from Fig. 2.2 was filtered with the function $\psi = X_2$ instead of the optimum filter ψ_{opt}, the resulting error was as shown in Fig. 3.3.

The extreme right-hand column of the table in Fig. 3.3 corresponds to $|N_0 - N_1|$ and represents the increase in error resulting from switching the output value for that particular input. It is also known as the *advantage*.[1]

For the two filters described, their outputs differ only for inputs (0, 1, 1) and (1, 0, 0) and the error therefore increases by nine and thirteen pixels respectively, giving a total increase of 22. The overall error would therefore be the error from the optimal filter plus the increase in error using the suboptimal filter (i.e., $14 + 22 = 36$ pixels). It can be seen from the extreme right-hand column of the table in Fig. 3.3 that the consequences of getting the filter output wrong can be very different for different inputs. Switching the filter output values for some inputs may have little effect on the MAE because either those inputs are not seen very often or the difference between N_0 and N_1 is very small.

It is interesting to observe that given no other information, the total number of pixels in error in the noisy image prior to filtering I_n may also be determined from

Num	X_2	X_1	X_0	N_0	N_1	OPT F	Sub F	Extra Error
0	0	0	0	64344	3	0	0	64341
1	0	0	1	18	1	0	0	17
2	0	1	0	10	1	0	0	9
3	0	1	1	2	11	1	0	9
4	1	0	0	16	3	0	1	13
5	1	0	1	1	4	1	1	3
6	1	1	0	1	12	1	1	11
7	1	1	1	2	87	1	1	85
							Total	22

Figure 3.3 A comparison of filters. The column on the right shows the increase in error resulting from the use of a filter with a different output for each input. In this case, the filters differ only for inputs (0,1,1) and (1,0,0).

Figure 3.4 Original example. The original example from Fig. 2.1 has 10% additive noise. The number of pixels in error is 5954, which corresponds to a MAE of 9%.

the observations table. This may be calculated by setting the filter function $\psi = X_1$ (i.e., the identity or "do nothing" filter) instead of the optimum. The error of 26 pixels may then be calculated by summing N_1 when $\psi = X_1 = 0$ and N_0 when $\psi = X_1 = 1$.

Having described the method on a simple example, it will now be applied to the original image shown in Fig. 2.1. The ideal image and the noise-corrupted version with 10% additive noise are shown in Fig. 3.4. The total number of pixels in error is 5954 which is a MAE of 9%.

Before proceeding to optimum image filtering, it is interesting to apply the median filter. This will be applied within a 3 × 3 window. The median reduces the pixels in error to 468 (0.71%) but a repeated application has little further effect reducing the error to only 443 pixels (0.67%). The corresponding images are shown in Fig. 3.5. The optimum filter was designed within a 3 × 3 window using the procedure described in Chapter 2. It was then applied to the noisy image reduc-

Applying a 3 × 3 median

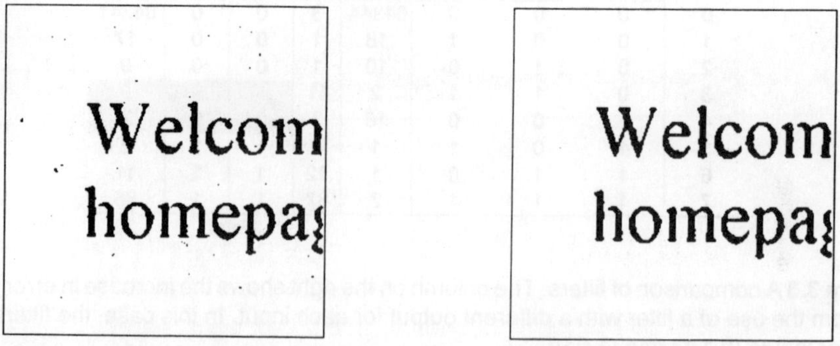

Once
468 pixels in error: 0.71%

Twice
443 in error: 0.67%

Figure 3.5 Median filtering in a 3 × 3 window. The median filter reduces the error on the first pass but has little further effect on the second. While much of the noise is removed, the text is damaged.

Optimum restoration window

Using 3 × 3 window

Using 5 × 5 window

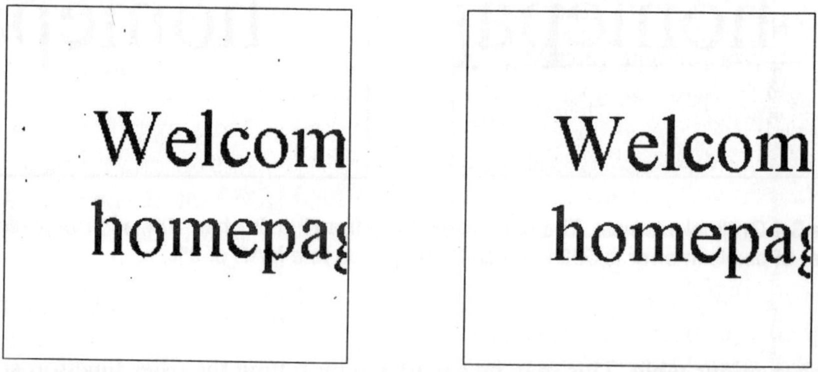

Error = 144 pixels: 0.22%

Error = 23 pixels: 0.03%

Figure 3.6 Optimum filtering. The results of applying the optimum filter within a 3 × 3 and 5 × 5 window are shown. They result in much lower errors than the median filters. The damage to the text is much less than for the median filters in Fig. 3.5.

ing the error to 144 pixels. For comparison, the optimum filter in a 5 × 5 window was also designed and applied. It had an error of just 23 pixels. The results of filtering with these optimum filters are shown in Fig. 3.6.

In order to investigate why the optimum filter is so much better than the median, the filters implemented in a five-point cross will be analyzed. The results of these filters are shown in Fig. 3.7. The optimum filter ψ_{opt} gives an error of 360 pix-

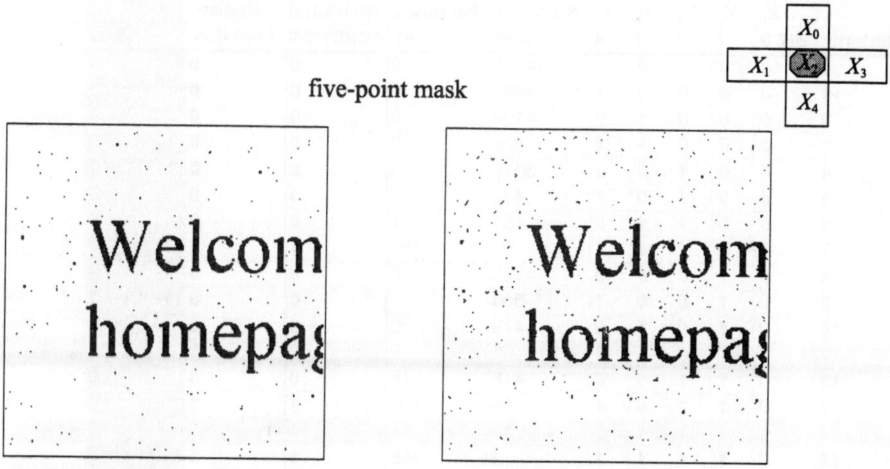

Figure 3.7 Comparison of optimum and median filters. The above filters implemented within a five-point window are shown. The optimum filter performed more than twice as well as the equivalent median.

els, whereas the median filter ψ_{med} has over twice the error with 754 pixels. Figure 3.8 shows the observation table, which has 32 input combinations (= 2^5). The difference in the error of the two filters of 394 pixels can be seen to correspond to the sum of the advantages for the inputs where the two filters differ. It is interesting to note that for the inputs where the two filters differ, $\psi_{opt} = 0$ and $\psi_{med} = 1$ and never the other way around. Also for these inputs, the value of the pixel at the center of the window $X_2 = 0$. The trained filter has therefore only learned to switch some of the black pixels (= 1) to white (= 0). The noise in the training set was purely additive and therefore the behavior of the correcting filter is subtractive. The median, on the other hand, treats black and white pixels equally and would give the same result if the input image were to be inverted, filtered, and re-inverted. The median filter is therefore unsuitable for correcting noise processes other than those that are symmetrical, i.e., are both additive and subtractive by equal proportions.

The properties of the median and its variants are discussed in more detail in Chapter 6.

3.2 Other Applications

3.2.1 Edge noise

The examples up to this point have focused on the removal of salt-and-pepper noise. Some further examples will now be given. An interesting case is edge noise.

	X_0	X_1	X_2	X_3	X_4	No times	No times	OPTIMUM	Median	
Minterm	BIT 0	1	2	3	4	X=0	X=1	FUNCTION	Function	Adv
0	0	0	0	0	0	34601	0	0	0	0
1	0	0	0	0	1	4002	0	0	0	0
2	0	0	0	1	0	4102	0	0	0	0
3	0	0	0	1	1	573	0	0	0	0
4	0	0	1	0	0	3817	3	0	0	0
5	0	0	1	0	1	433	0	0	0	0
6	0	0	1	1	0	451	14	0	0	0
7	0	0	1	1	1	61	110	1	1	0
8	0	1	0	0	0	4123	0	0	0	0
9	0	1	0	0	1	566	0	0	0	0
10	0	1	0	1	0	479	0	0	0	0
11	0	1	0	1	1	81	0	0	1	81 ←
12	0	1	1	0	0	448	18	0	0	0
13	0	1	1	0	1	65	106	1	1	0
14	0	1	1	1	0	56	44	0	1	12 ←
15	0	1	1	1	1	2	246	1	1	0
16	1	0	0	0	0	4003	0	0	0	0
17	1	0	0	0	1	446	0	0	0	0
18	1	0	0	1	0	585	0	0	0	0
19	1	0	0	1	1	65	0	0	1	65 ←
20	1	0	1	0	0	423	2	0	0	0
21	1	0	1	0	1	48	1	0	1	47 ←
22	1	0	1	1	0	65	114	1	1	0
23	1	0	1	1	1	5	469	1	1	0
24	1	1	0	0	0	557	0	0	0	0
25	1	1	0	0	1	90	0	0	1	90 ←
26	1	1	0	1	0	86	0	0	1	86 ←
27	1	1	0	1	1	13	0	0	1	13 ←
28	1	1	1	0	0	69	123	1	1	0
29	1	1	1	0	1	5	455	1	1	0
30	1	1	1	1	0	5	222	1	1	0
31	1	1	1	1	1	1	2263	1	1	0
							Median	Cost		394
							Error	Optimum		360
							Total	Error		754

Figure 3.8 Comparison of optimum and median filters.

When a picture is scanned, there can be errors at the edges. An example of this is shown in Fig. 3.9. In this case, the noisy image has an error of only 168 pixels so the MAE is tiny, however the effects are eye-catching and make the edges of the objects look "furry". A median filter makes little impression reducing the error to 153 pixels. A further iteration has little effect (150-pixel error). The optimum filter in a 5×5 window, however, reduces the error to just 34 pixels. It would be difficult to "guess" a filter that would perform close to the optimum for this type of example.

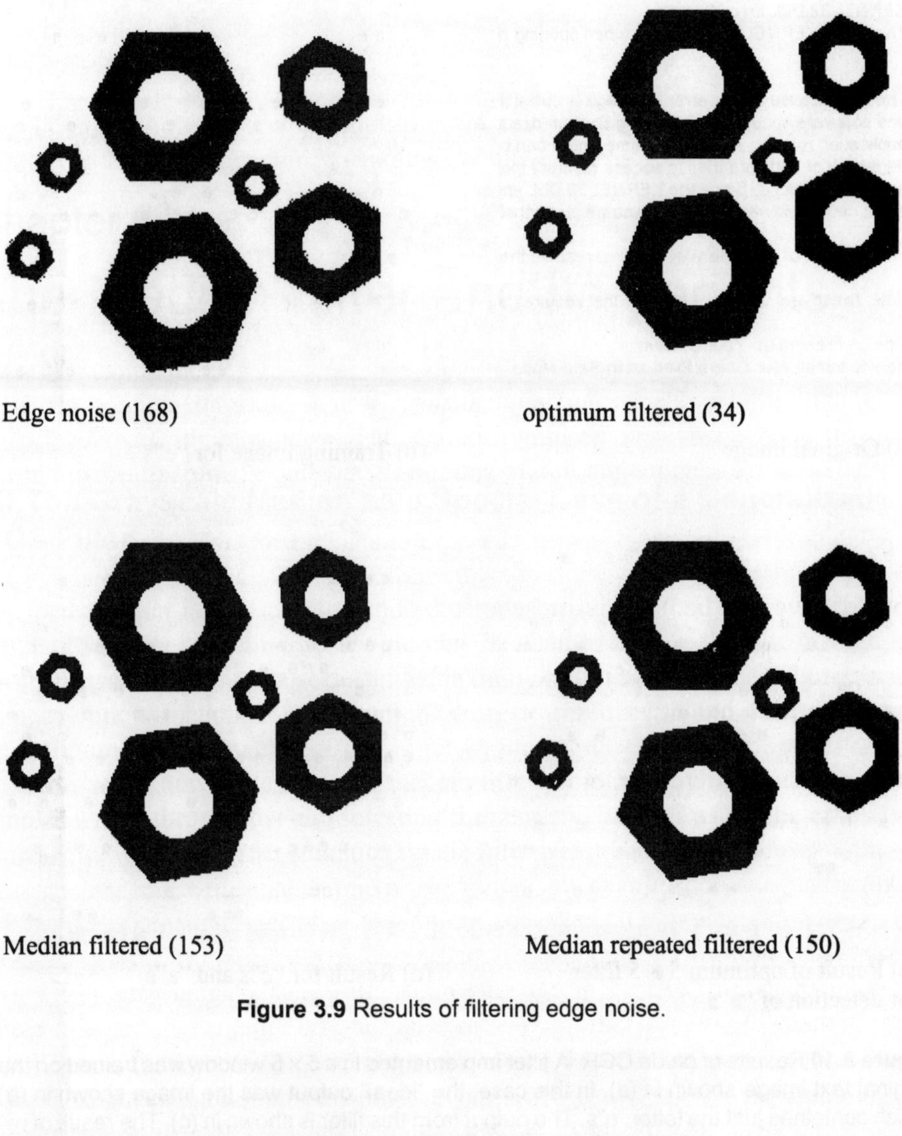

Edge noise (168) optimum filtered (34)

Median filtered (153) Median repeated filtered (150)

Figure 3.9 Results of filtering edge noise.

3.2.2 Simple optical character recognition

The techniques described may not only be used for removal of different types of noise from images, but also for recognition. This means that they may be trained to admit certain structures found in the images and to reject others. To demonstrate this property, a filter implemented within a 5 × 5 window was trained on the page of text shown in Fig. 3.10(a). In this case, the "ideal" output corresponds to the image shown in Fig. 3.10(b) containing only the letter "e"s from the original text. The output from this filter is shown in Fig. 3.10(c). The results of repeating the process for letters "a" and "e" are shown in Fig. 3.10(d). It can be seen that the recognition of the characters is quite accurate. It should be remembered that the algorithm has no

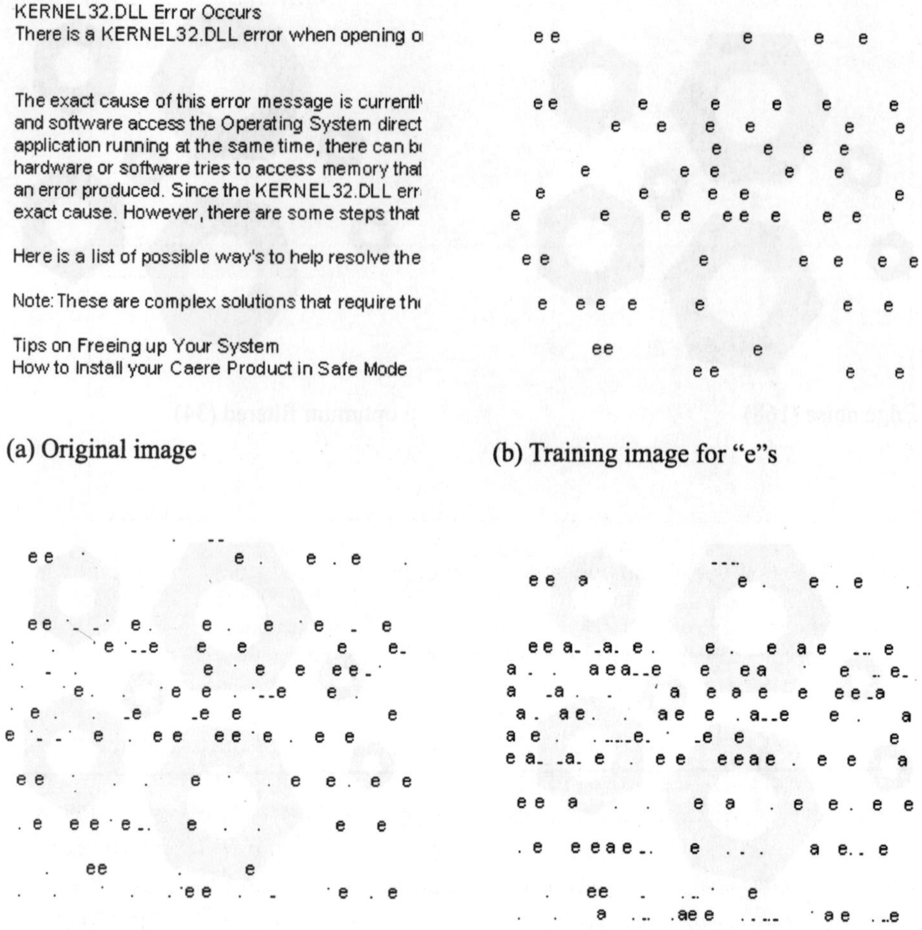

(a) Original image

(b) Training image for "e"s

(c) Result of optimum 5 × 5 filter for detection of "e"s

(d) Result for "e"s and "a"s

Figure 3.10 Results of crude OCR. A filter implemented in a 5 × 5 window was trained on the original text image shown in (a). In this case, the "ideal" output was the image shown in (b) which contained just the letter "e"s. The output from this filter is shown in (c). The result of repeating the process for letters "a" and "e" is shown in (d).

knowledge of the alphabet other than the training process and the resulting images could easily be cleaned up further by post processing, leaving just the characters.

3.2.3 Resolution conversion

An important problem in image processing that is often overlooked is that of resolution conversion. There is little point in having a 600 or 1200 dpi printer if the

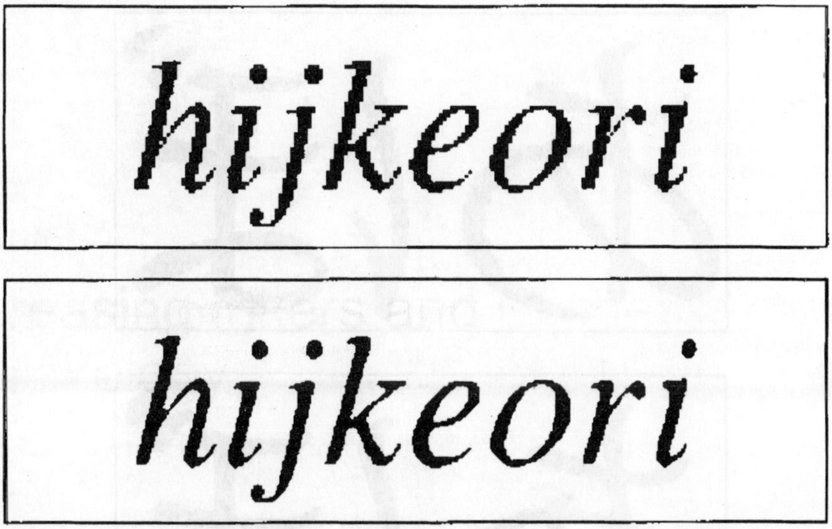

Figure 3.11 Results of resolution conversion. The upper image was originally scanned at 300 dpi. The lower image shows the results of filtering by a 17-point window trained on hundreds of images to convert from 300 to 600 dpi. A hardware implementation takes less than 100 gates. Reproduced from Loce[2] where further details may be found.

original document was scanned at 300 dpi or if the resolution of the text font is low. Filters of the type described above may be trained to carry out resolution conversion from lower to higher resolution. An example of resolution conversion taken from Loce is shown in Fig. 3.11.[2] A 17-point window was trained on hundreds of examples of low- to high-resolution images. The example in the figure was implemented in hardware with fewer than 100 gates. It can be seen that the upper image scanned at 300 dpi has very jagged edges. After filtering, the resolution was converted to 600 dpi and the edges are much smoother.

In order to test the robustness of the approach, a filter trained on a standard western alphanumeric font was applied to kanji characters shown in Fig. 3.12. It can be seen that the conversion works equally well even though these characters had not been seen in training. This suggests that the image statistics for many types of characters are very similar. In the above examples, there is no linear equivalent filter to solve these problems.

3.3 Summary

This chapter has shown how the table of observations may be used for a number of tasks related to filter error. In particular it has shown how the error after filtering with the optimum filter may be calculated. It has also shown how the error between filters may be compared. Other properties of the filter such as whether they are additive or subtractive may also be deduced. Lastly, a number of practical examples

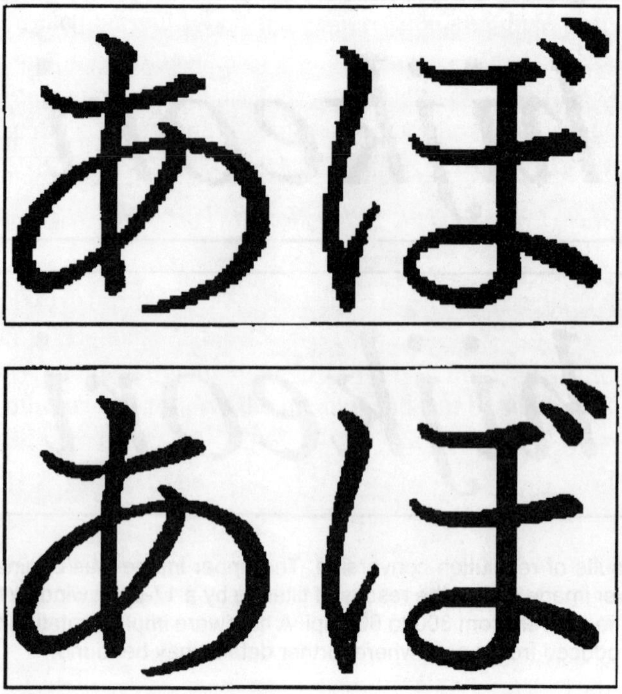

Figure 3.12 Demonstration of resolution conversion robustness. The upper image is a kanji character originally scanned at 300 dpi. It has been filtered by a 5 × 5 window trained on western alphanumeric characters. The accurate results suggest that the statistics of different alphabets are very similar. Reproduced from Loce[2] where further details may be found.

of filtering binary images for other applications have been presented. In all cases, the filters were derived from a training set and applied to data from that set. In practice, the process of filter training is more complex and this is discussed in detail in the next chapter.

References

1 E. R. Dougherty and J. Barrera, "Logical Image Operators," in *Nonlinear Filters for Image Processing*, E. R. Dougherty and J. Astola (eds.), SPIE Press, Bellingham, WA, 1–60 (1999).

2 R. P. Loce and E. R. Dougherty, *Enhancement and Restoration of Digital Documents: Statistical Design of Nonlinear Algorithms*, SPIE Press, Bellingham, WA (1997).

Chapter 4

How Do You Train the Filter for a Task?

At this stage, the reader might be asking the obvious question of why do we need to restore an image if the ideal original is available? In practice, a filter is used that has been designed on a representative training set. This means that examples similar to the image to be restored must be produced in some way. In the case of a fax machine, this is easy—a test image would simply be passed through the same process. The same is true for resolution changing and OCR examples discussed in the previous chapter. For old film restoration and other processes, it can be more difficult to recreate an ideal image for training. A section at the end of this chapter deals with this subject in more detail.

The extent to which a filter trained on one image may be applied to another is known as its *robustness*. If the statistics of either the noise or the image content of

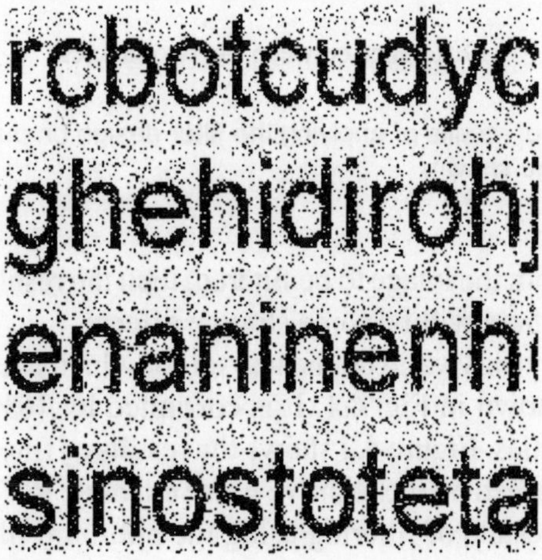

Figure 4.1 Test image with 10% additive and subtractive noise (6393 pixel error).

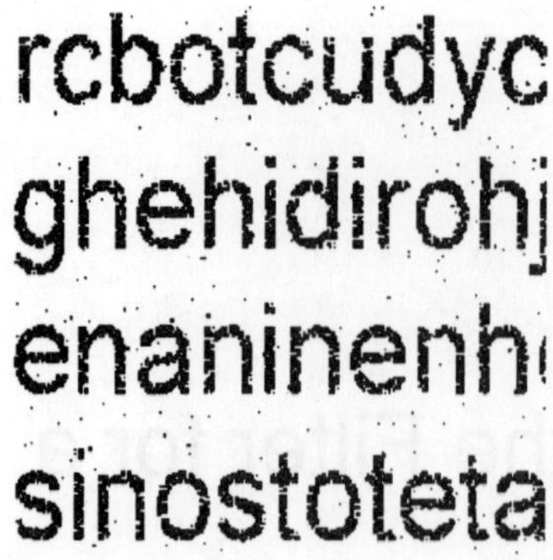

Result of applying a 5-point filter trained on additive noise only to image containing both additive and subtractive noise (2436 pixel error).

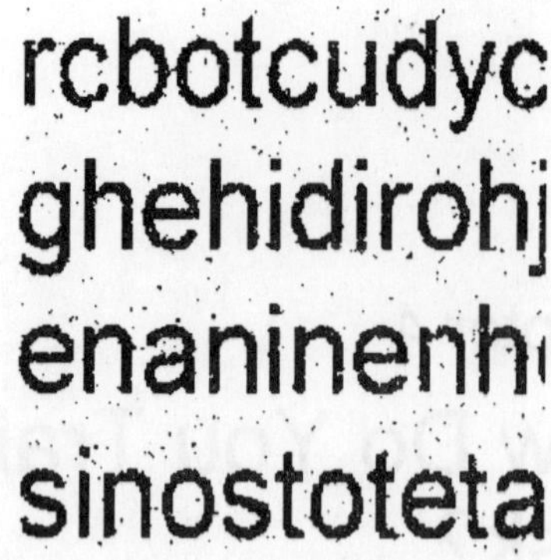

Result of applying optimal a 5-point filter to image containing both additive and subtractive noise (1462 pixel error).

Difference between the two images

Figure 4.2 Comparison of results of applying the filter design on purely additive noise with optimal filter for the image shown.

the training set vary from the image to be filtered, then the results will be suboptimal. In the extreme cases, there will be some strange effects.

A filter is generally only valid on the test set for which it has been trained. Consider the test image shown in Fig. 4.1. It is corrupted by 10% noise with additive and subtractive properties. Figure 4.2 shows the results of filtering this image using the filter previously trained on the additive noise example in Chapter 3. A comparison is made with the results of filtering using the optimum filter for this image.

The optimum filter reduces the number of pixels in error from 6393 to 1462, whereas the additive noise filter leaves 2436 pixels still in error. The nature of the

NUM	N_0	N_1	OPT F	Error F	SUBF	Error	Diff
0	26437	2	0	2	0	2	0
1	3430	9	0	9	0	9	0
2	3840	9	0	9	0	9	0
3	676	37	0	37	0	37	0
4	2929	7	0	7	0	7	0
5	379	70	0	70	0	70	0
6	416	61	0	61	0	61	0
7	78	400	1	78	1	78	0
8	3900	7	0	7	0	7	0
9	610	48	0	48	0	48	0
10	543	24	0	24	0	24	0
11	99	104	1	99	0	104	5
12	435	68	0	68	0	68	0
13	76	403	1	76	1	76	0
14	65	171	1	65	0	171	106 ←
15	7	1052	1	7	1	7	0
16	3397	6	0	6	0	6	0
17	440	30	0	30	0	30	0
18	644	44	0	44	0	44	0
19	99	149	1	99	0	149	50 ←
20	383	81	0	81	0	81	0
21	53	245	1	53	0	245	192 ←
22	95	405	1	95	1	95	0
23	9	1487	1	9	1	9	0
24	644	47	0	47	0	47	0
25	114	171	1	114	0	171	57 ←
26	113	109	0	109	0	109	0
27	19	583	1	19	0	583	564 ←
28	73	407	1	73	1	73	0
29	4	1540	1	4	1	4	0
30	10	1011	1	10	1	10	0
31	2	5710	1	2	1	2	0
				1462		2436	974

Figure 4.3 Table of observations showing filter functions, error for the optimum filter, and the earlier filter designed on purely additive noise.

remaining error is interesting. The filter trained on the earlier image with only additive noise matches the optimum filter in removing the noisy black pixels in the background. It is, however, unable to repair any of the noisy white pixels on the black text. It has also allowed the edges of the text to be thinned. The difference between the two images is shown in Fig. 4.2. The observation table and comparison between the filters is given in Fig. 4.3. The errors may be easily computed as in the previous chapters.

4.1 Effect of Window Size

As seen in the examples of the previous chapter, the error resulting from filtering with the optimum filter in any given window reduces with an increase in window size. For an example of practical noise, the window size for acceptable results may need to be 5 × 5 or larger.

Figure 4.4 shows the error resulting from filtering an image with the optimum filter in windows of increasing size.

It can be seen that the error declines exponentially as the window size increases. Why does this happen? The filter is, in effect, an estimator that attempts to determine the "true" value of the pixel in the ideal image. Depending on the statistics of the image, the larger the window, the more information the filter has to make a decision. From the values in the observation table, it is possible to compare two sub-windows and also to determine the increase in error caused by reducing the size of the window.

Consider the effect when instead of filtering with the 5-point window shown in Fig. 4.5(a), the filtering takes place within the 3-point asymmetrical window formed by omitting pixels X_3 and X_4. This is shown in Fig. 4.5(b) and the intention is still to estimate the true value of the pixel at location X_2.

The observation table generated for the original 5-point window is shown in Fig. 4.6(a). The observation table for the 3-point window is calculated by combining all inputs with the same value of pixels for X_0, X_1, and X_2 regardless of the val-

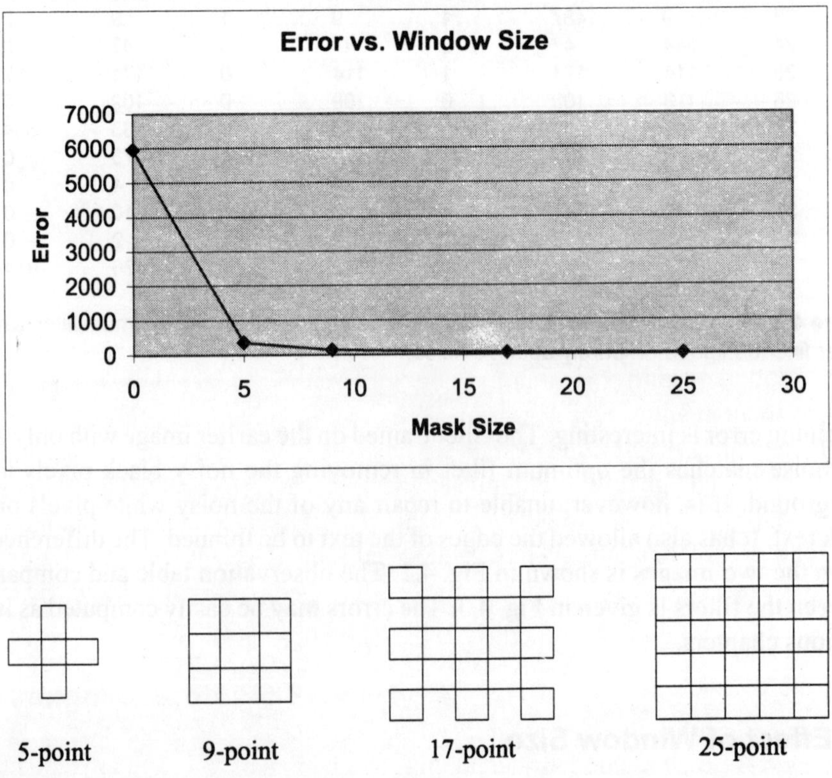

Figure 4.4 Effect of window size on MAE for optimum filter. In all cases the error falls with increasing window size.

Figure 4.5 Two different filter windows. (a) The 5-point cross and (b) the 3-point asymmetrical window formed by omitting pixels X_3 and X_4.

ues of X_3 and X_4. This is carried out by summing N_0 and N_1 for these inputs. The net effect is that a single output must be allocated to each combination of X_0, X_1, and X_2. The new table of observations shown in Fig. 4.6(b) has only eight inputs and the error is now 3006 pixels.

Figure 4.7 shows the source of the errors. Effectively each set of four separate inputs for the original 5-point window must now all have the same output for the 3-point sub-window. Those outputs that differ from the new combined value result in an increase in error. The total increase in error is 1544, which corresponds precisely to the difference between the error for the 5-point window (1462) and the error for the 3-point window (3006).

This method may be used to compare any two sub-windows within an overall region of support by omitting certain inputs. In the example shown, these were the two least significant variables, so the inputs to be combined were adjacent in the table. For other inputs the table must be rearranged. Also, it is possible to use this technique to compare windows at different resolutions though this is beyond the scope of this book. For details see Dougherty et al.[1]

From all the evidence thus far, it would therefore seem that it is better to use a large window. This is true provided that the optimum filter for the large window can be found. This task, however, gets progressively harder as the size of the window increases.

In Chapter 2 it was shown that a 3-point window had $2^3 = 8$ input combinations and $2^{2^3} = 256$ possible functions. Therefore for a window with n points, there are 2^n input combinations and hence 2^{2^n} functions. Consider Table 4.1. The number of input combinations and associated functions scale at an alarming rate. Even the simple 5-point cross used in Chapter 3 is capable of implementing more than 4.29×10^9 functions! For window sizes of 17 and 25 points, the number of filter functions is too large to express in terms of standard floating-point numbers.

So what effect does this rapid increase in the number of possible filters cause when designing an optimum filter? The key column in the table is the number of input combinations possible. Recalling the design process in Chapter 2, a table of observations was constructed from the training set. The size of this table corresponds to the number of input combinations. For each input it is necessary to observe its occurrence a sufficient number of times to make a good statistical estimate of the optimum

	X_0	X_1	X_2	X_3	X_4	N_0	N_1	N_0	N_1	F3
0	0	0	0	0	0	26437	2	33704	47	0
1	0	0	0	0	1	3430	9			
2	0	0	0	1	0	3397	6			
3	0	0	0	1	1	440	30			
4	0	0	1	0	0	3840	9	5259	239	0
5	0	0	1	0	1	676	37			
6	0	0	1	1	0	644	44			
7	0	0	1	1	1	99	149			
8	0	1	0	0	0	2929	7	3744	403	0
9	0	1	0	0	1	379	70			
10	0	1	0	1	0	383	81			
11	0	1	0	1	1	53	245			
12	0	1	1	0	0	416	61	598	2353	1
13	0	1	1	0	1	78	400			
14	0	1	1	1	0	95	405			
15	0	1	1	1	1	9	1487			
16	1	0	0	0	0	3900	7	5268	273	0
17	1	0	0	0	1	610	48			
18	1	0	0	1	0	644	47			
19	1	0	0	1	1	114	171			
20	1	0	1	0	0	543	24	774	820	1
21	1	0	1	0	1	99	104			
22	1	0	1	1	0	113	109			
23	1	0	1	1	1	19	583			
24	1	1	0	0	0	435	68	588	2418	1
25	1	1	0	0	1	76	403			
26	1	1	0	1	0	73	407			
27	1	1	0	1	1	4	1540			
28	1	1	1	0	0	65	171	84	7944	1
29	1	1	1	0	1	7	1052			
30	1	1	1	1	0	10	1011			
31	1	1	1	1	1	2	5710			

(a)

	X_0	X_1	X_2	N_0	N_1	F3	Error
0	0	0	0	33704	47	0	47
1	0	0	1	5259	239	0	239
2	0	1	0	3744	403	0	403
3	0	1	1	598	2353	1	598
4	1	0	0	5268	273	0	273
5	1	0	1	774	820	1	774
6	1	1	0	588	2418	1	588
7	1	1	1	84	7944	1	84
					Total		3006

(b)

Figure 4.6 Two tables of observations. (a) gives the observations within the 5-point window. (b) shows the observations within the 3-point window formed by omitting pixels X_3 and X_4. The table shown in (b) is formed by combining lines in original table (a) that have the same values of X_0, X_1 and X_2.

	X_0	X_1	X_2	X_3	X_4	N_0	N_1	F5	F3	
0	0	0	0	0	0	26437	2	0	0	
1	0	0	0	0	1	3430	9	0		
2	0	0	0	1	0	3397	6	0		
3	0	0	0	1	1	440	30	0		
4	0	0	1	0	0	3840	9	0	0	
5	0	0	1	0	1	676	37	0		
6	0	0	1	1	0	644	44	0		
7	0	0	1	1	1	99	149	1		50
8	0	1	0	0	0	2929	7	0	0	
9	0	1	0	0	1	379	70	0		
10	0	1	0	1	0	383	81	0		
11	0	1	0	1	1	53	245	1		192
12	0	1	1	0	0	416	61	0	1	355
13	0	1	1	0	1	78	400	1		
14	0	1	1	1	0	95	405	1		
15	0	1	1	1	1	9	1487	1		
16	1	0	0	0	0	3900	7	0	0	
17	1	0	0	0	1	610	48	0		
18	1	0	0	1	0	644	47	0		
19	1	0	0	1	1	114	171	1		57
20	1	0	1	0	0	543	24	0	1	519
21	1	0	1	0	1	99	104	1		
22	1	0	1	1	0	113	109	0		4
23	1	0	1	1	1	19	583	1		
24	1	1	0	0	0	435	68	0	1	367
25	1	1	0	0	1	76	403	1		
26	1	1	0	1	0	73	407	1		
27	1	1	0	1	1	4	1540	1		
28	1	1	1	0	0	65	171	1	1	
29	1	1	1	0	1	7	1052	1		
30	1	1	1	1	0	10	1011	1		
31	1	1	1	1	1	2	5710	1		
										1544

Figure 4.7 Errors resulting from use of a sub-window. Each set of four inputs is combined to have a single output. An increase in error occurs when individual outputs differ from the new combined output.

Table 4.1 Increasing the size of the window. The table shows the number of lines in the table of observations and the number of possible functions for various sizes of window. It can be seen that these rise rapidly as the size of the window is increased.

Pixels	Combinations	Functions
n	2*n*	22*n*
3	8	256
5	32	4294967296
9	512	1.3408E + 154
17	131072	Too big to show
25	33554432	

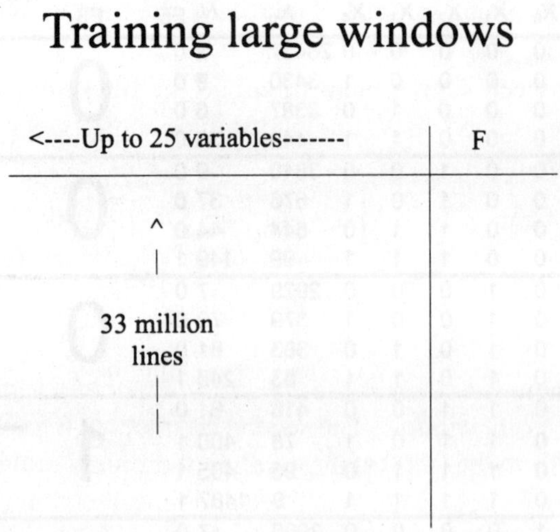

Figure 4.8 Problems with large window sizes. For larger window sizes, the number of lines in the table of observations very rapidly becomes extremely large. This means that even the training data from many thousands of images is spread very thinly throughout the table. Many inputs will not have been observed a sufficient number of times to make a statistically robust estimate. Even more inputs will not have been observed at all.

output value. In the examples given in previous chapters, every input was seen many times. However, as the size of the filter (and hence the table of observations) increases, the table can become very sparse. The observation table for a 25-point window contains over 33 million lines (see Fig. 4.8). If this were to be trained on a 512 × 512 image, there would only be a quarter of a million observations to distribute over 33 million lines, and hence most of the counts in the table would be zero. It can now be seen that the 17-point window is attractive. Despite spanning a similar region of support, it has only 131,072 possible inputs making it much easier to train on just a few images. Where a particular input is not seen in the training set, the filter does not know which value to allocate for its output. If that particular input is encountered in the actual image to be filtered, the output may be arbitrary leading to large errors.

4.2 Training Errors

Let it be assumed that ψ_{opt} is the optimum filter for a given task. If the filter ψ_n is the best filter that may be implemented within an n-point window, then ψ_n will be suboptimal to ψ_{opt} and hence ψ_n is a constrained version of ψ_{opt}.

If, however, the windowed filter is produced by training on a fixed number of samples N, the resulting filter $\psi_{n,N}$ will be further suboptimal to both ψ_n and ψ_{opt}.

$$MAE\left(\psi_{n,N}\right) \geq MAE\left(\psi_n\right) \geq MAE\left(\psi_{opt}\right) \tag{4.1}$$

As the number of training samples N increases, the trained filter becomes closer to the optimal, i.e., $\psi_{n,\infty} \rightarrow \psi_n$.

The error between the optimum filter and the filter implemented within an n-point window and trained on N training samples consists of two components.

$$E\left[\Delta\left\langle \psi_{n,N}, \psi_{\text{opt}}\right\rangle\right] = \Delta\left\langle \psi_n, \psi_{\text{opt}}\right\rangle + E\left[\Delta\left\langle \psi_{n,N}, \psi_n\right\rangle\right] \tag{4.2}$$

total error = constraint error + estimation error

The first component is known as the *constraint error* and is due to the filter being restricted to an n-point window. The second component is known as the estimation error and results from the fact that the number of training samples is finite. The constraint error is *deterministic*, i.e., it is fixed and repeatable for a given problem. The estimation error is *stochastic*. This means that it is a statistical quantity and will vary if the design process is repeated a number of times with different training data.

As has been seen in early examples, the *constraint error* reduces with increasing n. The bigger the window, the more accurate the filter.

The *estimation error* reduces with increased training as can be seen in Fig. 4.9(a).[1] Notice that the estimation error for smaller windows converges very rapidly. However for some of the larger windows, the convergence is very slow and even after 700,000 samples the 21-point window is showing a larger estimation error than the smaller windows did at the start. This error is because the filter is undertrained, i.e., the amount of training data is insufficient. The amount of data required to reduce the estimation error to a reasonable level may be impossibly large. When combined with the convergence error, the total error versus training data is shown in Fig. 4.9(b). The filters implemented in the smaller windows converge very quickly. The filters implemented in larger windows eventually converge to a lower error, but this can take a long time. For any given amount of data, a different window size might give the lowest error. For example, after 100,000 samples, the 9-point window gives the best filter but by 200,000 samples it has been superseded by the 13-point window. Eventually the 21-point window will give the lowest error, but this is still a long way away. In fact even after 700,000 samples, the results of filtering with the 21-point window are still worse than the original noisy image.

To illustrate this point, the results of Fig. 4.9 are presented differently in Fig. 4.10(a). The total error for any given filter is plotted against window size for fixed amounts of training data. For any size of training set, the error will fall to a minimum as the size of the window increases, after which it will rise very rapidly. Increasing the training set by an order of magnitude only serves to move the minimum to a slightly larger value of window size.

Depending on the problem, a smaller window might be sufficient. In the case of the graph in Fig. 4.10(b) the corrupting process was 5% salt-and-pepper noise. A small window size (5 points) was capable of removing much of the noise and

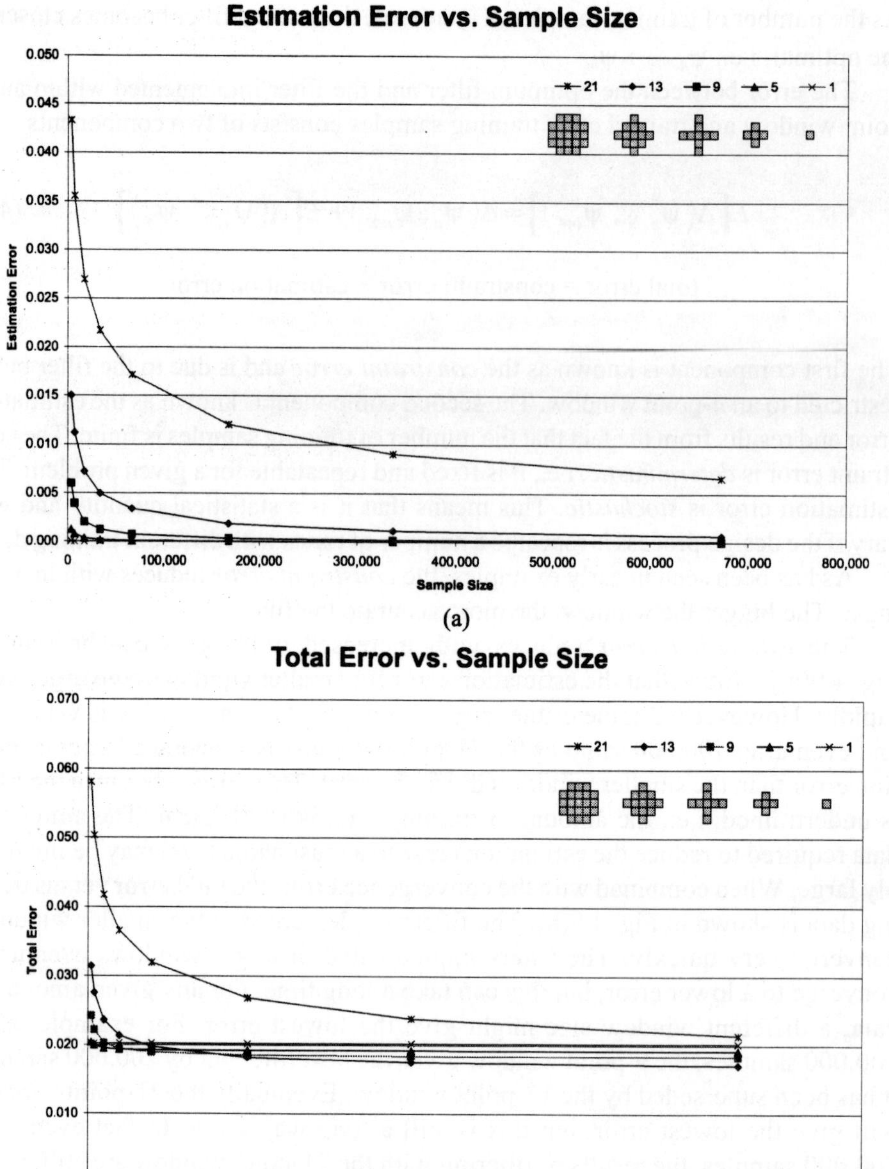

Figure 4.9 Estimation (a) and total error (b) for edge-noise problem. Problems with large window sizes[1]: For larger window sizes, massive amounts of training data are required to reduce the estimation error. It can be seen that even after 700,000 training samples, the larger windows still give a worse result than the original. (Reproduced from Ref. 1 with permission of Springer Science and Business Media.)

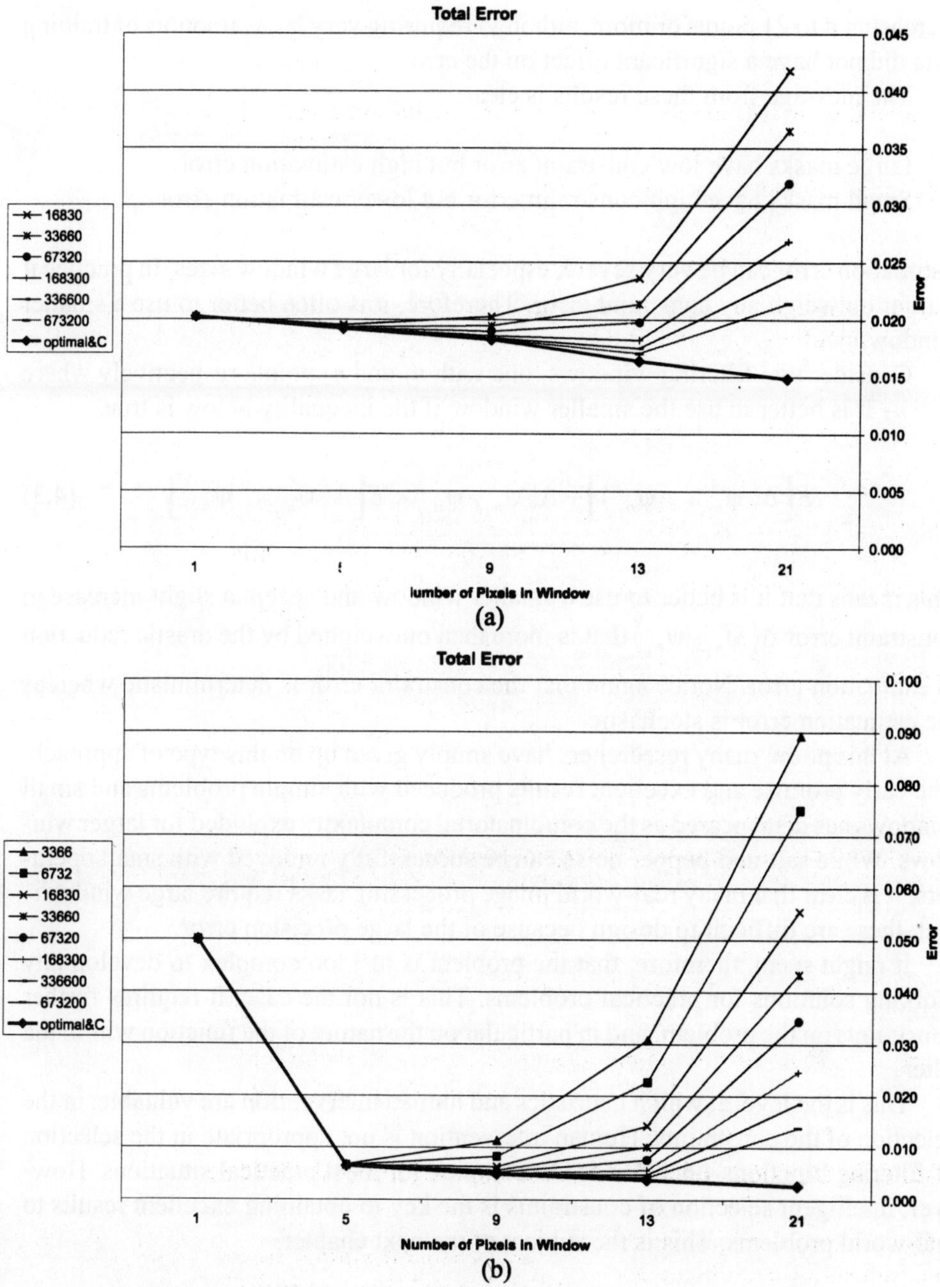

Figure 4.10 Problems with large window sizes: The plot shows the total error versus window size for fixed amounts of training data for (a) edge noise and (b) salt-and-pepper noise. It can be seen that large amounts of extra data are required for the larger windows. In the case of salt-and-pepper noise, the overall error does not decrease very much, even for bigger windows. (Reproduced from Ref. 1 with permission of Springer Science and Business Media.)

increasing it to 21 points or more, although requiring very large amounts of training data did not have a significant effect on the error.

The message from these results is clear:

- Large masks have low constraint error but high estimation error.
- Small masks have high constraint error but lower estimation error.

Estimation error can be very severe, especially for large window sizes. In practice it can far outweigh any constraint error. Therefore, it is often better to use a smaller window.

Consider two filtering windows, one with n_1 and n_2 points respectively where $n_1 > n_2$. It is better to use the smaller window if the inequality below is true.

$$E\left[\Delta\left\langle \psi_{n_2,N}, \psi_{n_2}\right\rangle\right] + \Delta\left\langle \psi_{n_2}, \psi_{n_1}\right\rangle < E\left[\Delta\left\langle \psi_{n_1,N}, \psi_{n_1}\right\rangle\right] \tag{4.3}$$

This means that it is better to use a smaller window and accept a slight increase in constraint error $\Delta\left\langle \psi_{n_2}, \psi_{n_1}\right\rangle$ that is more than outweighed by the drastic reduction in estimation error. Notice again that the constraint error is deterministic whereas the estimation error is stochastic.

At this point, many researchers have simply given up on this type of approach. The early promise and excellent results produced with simple problems and small windows has disappeared as the combinatorial complexity exploded for larger windows. While salt-and-pepper noise can be successfully removed with small operators, it is clear that many real-world image processing tasks require large windows. Yet, these are difficult to design because of the large precision error.

It might seem, therefore, that the problem is just too complex to develop any working solutions for practical problems. This is not the case; it requires further constraints on the problem, and in particular on the nature of the function within the filter.

This is the level at which heuristics and human intervention are valuable, in the selection of the constraints. Human intervention is not appropriate in the selection of filtering functions since this is too complex for most practical situations. However, intelligent selection of constraints is the key to obtaining excellent results to real-world problems. This is the subject of the next chapter.

4.3 In Defense of Training Set Approaches

A criticism sometimes leveled at these filter design methods is that a representative training set is required. This means that the "ideal" version of an image is required in order to restore the noisy version. While this is a valid criticism, it is unreasonable to dismiss such approaches simply because they make use of a test set.

There are several ways in which a test set can be made available. For example:

1. In practice it is often possible to duplicate a corruption process such as one resulting from a printing operation or fax transmission. A known test image may then be passed through the process and used to train a filter for use on other similarly corrupted images where the original is not available.
2. When presented with a noisy image or sequence to be restored, it is often possible to identify a clean part of the image and to cut and paste examples of noise corruption in order to create an ideal and noisy test set. This method has been used successfully in restoring old video sequences.[2–4]
3. An example of a corrupted image may also be cleaned manually using a software package such as Adobe Photoshop. This may then be used as the training set to design a filter for the automatic restoration of other images.

These methods may seem artificial, but in practice there may be few other options to solve real-world restoration problems. More theoretical approaches such as mathematical models can be used to simulate the statistical properties of images and noise processes. The optimum filters can then be found for these models. In practice the image and noise characteristics rarely conform to these simple models, especially in image and film corruption. Once the assumptions of the models are no longer valid, then the performance of such filters can rapidly decline.

An alternative method is to optimize an image quality criterion, and these do exist.[5] Most restoration approaches can be adapted to optimize such a criterion rather than to minimize the error with respect to a training set. In practice these methods have been found to lead to poorer results compared to training set approaches.[6]

4.4 Summary

This chapter has given insight into the problems associated with filter training. By definition, the filter must be trained on a different set of images than it is applied to in practice. The training set must be statistically consistent with the task in hand. This chapter has considered the effect of changing the size of the filter window and the associated implications for the size of the training set required. This chapter has also introduced the two types of error present in filters designed by training; namely constraint and estimation error. The criteria for whether or not the application of a constraint is beneficial have been quantified. Finally, an explanation of how training sets may be acquired for different classes of problems has been given. The next chapter considers one of the most commonly used forms of constraints, that of restricting the filter to increasing functions. This results in filters that have an interpretation in terms of mathematical morphology.

References

1 E. R. Dougherty, J. Barrera, G. Mozelle, S. Kim, and M. Brun, "Multiresolution analysis for optimal binary filters," *J. Math. Imaging Vis.*, **14**(1), 53–72 (2001).

2 N. R. Harvey and S. Marshall, "The use of genetic algorithms in morphological filter design," *Signal Processing: Image Communication*, **8**(1), 55–72 (1996).

3 N. R. Harvey and S. Marshall, "GA Optimisation of Multidimensional Gray-Scale Soft Morphological Filters with Applications in Archive Film Restoration, Mathematical Morphology and its Applications to Signal Processing," *ISMM 2000*, Palo Alto (2000).

4 M. R. Hamid, S. Marshall, and N. R. Harvey, "GA optimisation of multidimensional gray-scale soft morphological filters with applications in archive film restoration," *IEEE Trans. Circuits and Systems for Video Technology*, **13**(5), 406–416 (2003). See also **13**(7), 726 (2003).

5 G. Ramponi, N. Strobel, S. K. Mitra, and T-H. Yu, "Nonlinear unsharp masking methods for image contrast enhancement," *J. Electron. Imaging*, **5**(3), 353–366 (1996).

6 N. R. Harvey and S. Marshall, "Film Restoration using soft morphological filters," *Proceedings of the IEE 6th International Conference on Image Processing and its Applications IPA'97*, Dublin, Ireland (1997).

Chapter 5

Increasing Filters and Mathematical Morphology

5.1 Constraints on the Filter Function

In the previous chapter, it was seen that the estimation error of the filters increased rapidly with window size. This was because the function defining the behavior of the filter was unconstrained. Referring back to the design process described in Chapter 2, every line of the table of observations was treated as a separate independent entity. It was therefore necessary to see a sufficient number of examples of every possible input in order to design the filter. For small windows this was feasible. However, for larger windows the number of inputs was huge and it was impossible to see all of them. In practice it is not necessary for a filter to see all possible inputs in order to determine the function accurately. This means that an output value must be assigned to an input pattern that was never seen in training.

Consider the inputs shown in Fig. 5.1. Two of these input patterns were seen in the training set a sufficient number of times for the output to be allocated a value of 1. The other input patterns were never seen at all and in theory their output is unknown. However, it can be observed that the unknown patterns sit *between* the other two patterns and there is no reason to believe that their value should be anything other than 1. In the same way that a linear function may be interpolated with models such as spline functions, a logic function may be interpolated such that it fits the data at the known points and provides a good approximation at the undefined points. A common approach is to limit the filter to a particular type of function known as an *increasing* function. An increasing function is one that can be expressed without the use of negation, i.e.,

$$F_{\text{inc}}(\mathbf{x}) = X_1 + X_2 X_3 + X_4 X_5 X_6 \quad \text{is an increasing function} \tag{5.1}$$

$$G_{\text{non inc}}(\mathbf{x}) = X_1 + X_2 \overline{X}_3 + X_4 X_5 X_6 \quad \text{is a nonincreasing function.} \tag{5.2}$$

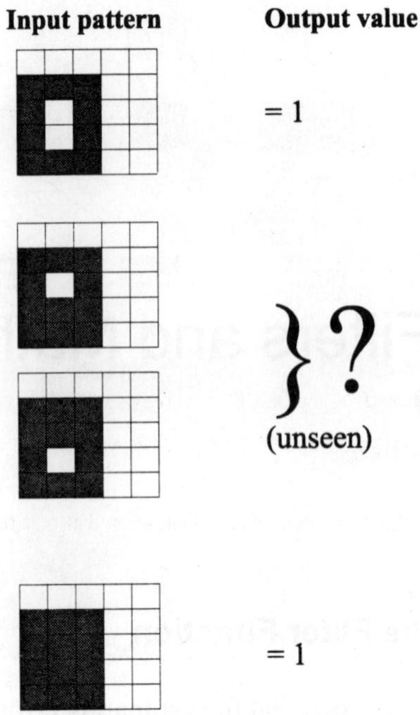

Figure 5.1 Assignment of output to unknown input patterns. If the patterns at the top and bottom are assigned a value of 1, it is reasonable to assume that the middle patterns should also be assigned a value of 1.

Increasingness implies a partial *ordering* of the input values of a function. That is,

$$F(\mathbf{x}) \geq F(\mathbf{y}) \quad \text{for} \quad \mathbf{x} \geq \mathbf{y}, \tag{5.3}$$

where $\mathbf{x} \geq \mathbf{y}$ implies that $X_i \geq Y_i$ for every component of **x** and **y**. For example, if $\mathbf{x} = (011)$ and $\mathbf{y} = (001)$ then $\mathbf{x} > \mathbf{y}$ and therefore it follows that $F(\mathbf{x}) \geq F(\mathbf{y})$ for any increasing function F. However for $\mathbf{x} = (010)$ and $\mathbf{y} = (001)$ there is no ordering of **x** and **y** and therefore nothing can be inferred about the ordering of $F(\mathbf{x})$ and $F(\mathbf{y})$.

A filter based on an *increasing* function is known as an *increasing* filter (ψ_{inc}). Increasingness is simply a further constraint on the filter. It will cause an increase in constraint error unless the optimal filter happens to be an increasing filter, i.e.,

$$\text{MAE}(\psi_{\text{opt}}) \leq \text{MAE}(\psi_{\text{inc}}). \tag{5.4}$$

Even though the best possible increasing filter may be inferior to the best filter overall, it will be easier to train because its search space will be significantly reduced. This means that the estimation error of the increasing filter will be much lower than for the filter without this constraint. The key to good filter design is to

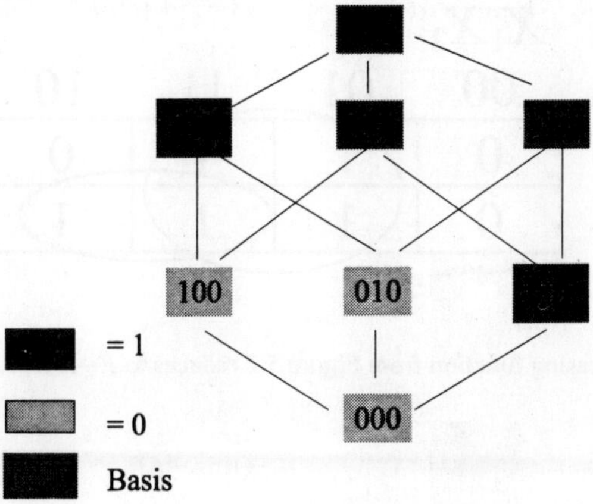

Figure 5.2 Lattice representation of a 3-input function.

determine a constraint that reduces the search space to allow training on a realistically sized training set but that allows sufficient flexibility to produce an accurate solution. As with filter window size, it is a trade off between estimation and constraint error. The amount by which the imposition of the increasingness constraint limits the filter should not be underestimated.

Consider the lattice representation of a function of three variables shown in Fig. 5.2.

The lattice has the value of $\mathbf{x} = (1, 1, 1)$ at the top and $\mathbf{x} = (0, 0, 0)$ at the bottom and all the values in between. The partial ordering is conveyed by the connecting lines, indicating that some values of $\mathbf{x}$ are above (or below) others, as defined by Eq. 5.3. This lattice structure can be extended to any number of variables, though it becomes increasingly complex to illustrate.

An increasing function causes the lattice to be cut into two sections—top and bottom. All of the inputs in the top section have a corresponding filter output of 1. All those in the bottom section have an output of 0. When an input is encountered for which the output is 1, then every input above that in the lattice can be assumed to have an output of 1. Similarly, when an input is found for which the output is 0, then every input beneath it has an output of 0. The entire function may be specified by identifying the minimum inputs for which the output is 1 as shown in Eq. 5.5. These inputs are known as the basis inputs.

$$\psi_{inc}(\mathbf{x}) = 1 \quad \text{if there exists } i \text{ such that} \quad \mathbf{x} \geq \mathbf{x}_{bi} \quad \text{and} \quad \psi_{inc}(\mathbf{x}) = 0$$
$$\text{for all} \quad \mathbf{x} < \mathbf{x}_{bi} \quad \text{for all } i. \tag{5.5}$$

The set of inputs $B[\psi_{inc}]$ is known as the basis. In Fig. 5.2 the basis contains two inputs $\mathbf{x}_{b1} = (1, 1, 0)$, $\mathbf{x}_{b2} = (0, 0, 1)$. This completely defines the function for all in-

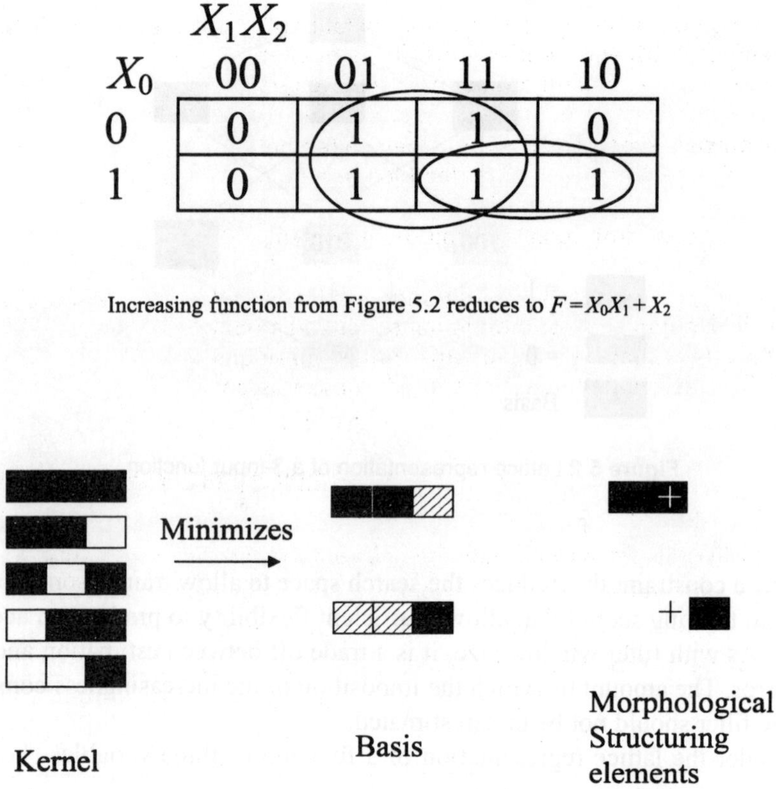

Figure 5.3 Minimization and implementation of an increasing function into morphological structuring elements.

puts. The design of an increasing filter can be reduced to the identification of the basis inputs that partition the set of all inputs into a lower and upper portion.

If this increasing filter ψ_{inc} is represented by the pixels within a 3-point horizontal window, then it may be implemented by determining if any of the basis inputs fits within the foreground of the image. This is equivalent to Boolean reduction where it is not necessary to test for every input of the filter, but it is sufficient to simplify the function and to test only for the reduced set.

Figure 5.3 shows how the increasing function may be reduced using the standard technique of a K map to produce the minimized function of

$$F = X_0X_1 + X_2 . \tag{5.6}$$

This means that a foreground (black) pixel will be produced if either X_2 is black OR both X_0 AND X_1 are black. If neither of these conditions holds, the pixel will be white. This function may therefore be implemented by testing if either of the sub-windows below fits the image.

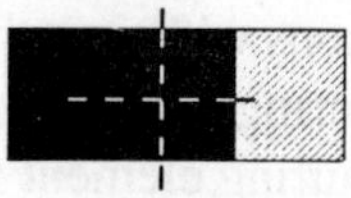

$X_0 X_1$

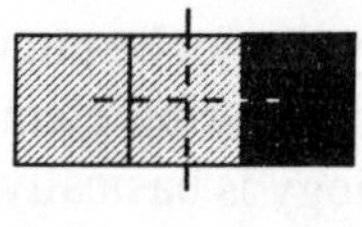

X_2

The hashed box indicates a don't care term.

This approach is directly equivalent to Mathematical morphology.[1,2,3] The shapes above that are tested against the image are in fact structuring elements.

In general, any logical function may be implemented through a sum of products expression. An increasing function is implemented when none of the variables are negated. Similarly in the world of mathematical morphology, any morphological operator may be written as a *union of erosions*.[4] These are in fact one and the same thing. The erosions are equivalent to the products (or the ANDs) and the union is equivalent to the sum. A number of different sub-components are tested against the image. If one or more of them fits the image, the overall result is true. In the morphological representation, a set of structuring elements are used. These are equivalent to the *minterms* in the logical representation. In set theory, this would be written as

$$\bigcup_{i \in B} I \Theta b_i, \tag{5.7}$$

where U represents the union operator, Θ is the erosion operator, I is the image, and b_i are the structuring elements equivalent to X_0X_1 and X_2 shown above.

In mathematical morphology literature, there are few clues to selecting the best set of structuring elements. In Soille's book of applications of morphological image processing, most of the structuring elements are designed heuristically (in other words, by guesswork).[5] There are some explanations in the literature but these are buried within other more involved texts.[6]

For comparison, an example of a 3-variable nonincreasing function, i.e., a function for which the increasingness property does not hold, is shown in Fig. 5.4.

The inputs $\mathbf{x} = (0, 0, 1)$ and $\mathbf{x} = (0, 1, 1)$ prevent this function from being increasing. It cannot be represented in the same way as the increasing function. The function may still be minimized using a K map as shown in Fig. 5.5.

$$F = X_0 X_1 + X_0 X_2 + \overline{X}_1 X_2 \tag{5.8}$$

The resulting basis inputs of the minimized function are shown below.

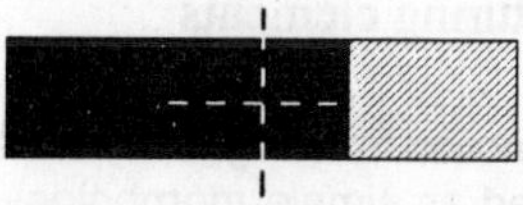

$X_0 X_1$

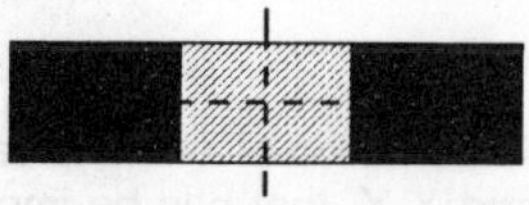

$X_0 X_2$

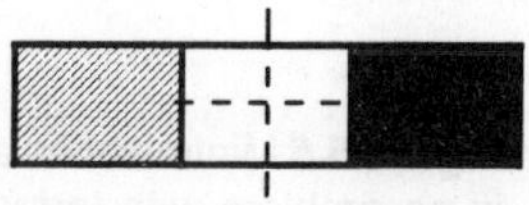

$\overline{X}_1 X_2$

They cannot be put into the context of a simple union of erosions with morphological structuring elements. This is because erosion by a structuring element in morphology is basically a foreground operation. Either the structuring element fits the

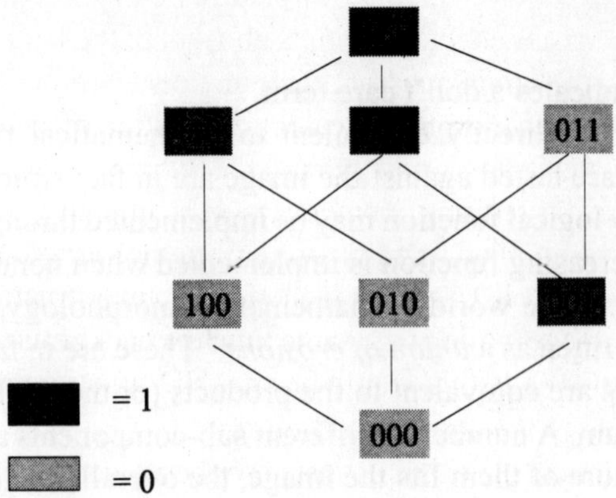

Figure 5.4 Lattice representation of a nonincreasing function.

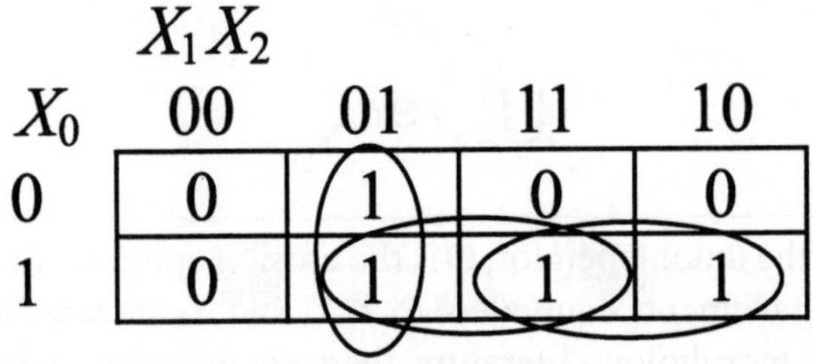

The nonincreasing function from Fig. 5.4 reduces to $F = X_0X_1 + X_0X_2 + \overline{X}_1X_2$.

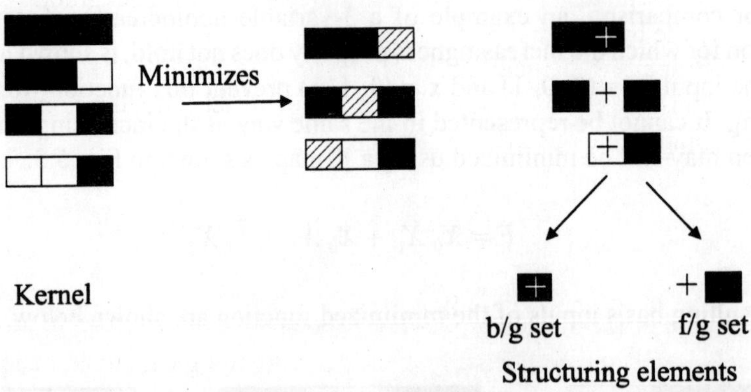

Figure 5.5 Minimization and implementation of a nonincreasing function from Fig. 5.4. There is no problem with terms X_0X_1 and X_1X_2 that can be implemented as simple morphology. However, term $\overline{X}_1X_2$ must be implemented as a hit-or-miss transform.

foreground or it does not. The background of the image is not considered. Consequently, the erosion may only model minterms that do not have negation. This limitation is related to the lattice structure representation of the increasing function. Once an input is found into which the structuring element will fit (i.e., its output is 1), it may be safely assumed that all inputs above it in the lattice also have an output of 1. However, for the nonincreasing function no such order may be assumed.

In order to produce a morphological representation of a nonincreasing function such as the one above, it is necessary to use the *hit-or-miss transform*.[7] In this operation, the kernel of structuring elements is split into two parts: foreground and background. They are linked in pairs—one from the foreground and one from the background. The output is true only if the foreground structuring element fits the foreground of the image while at the same time the background structuring element fits the background. While this would be an AND function in Boolean algebra, in set theory terminology it is an intersection $\cap$.

The only problem in the example just described is caused by the operator below, which must be decomposed into a foreground and background element as shown.

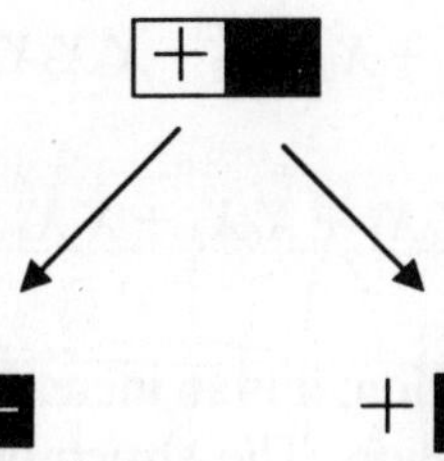

Background SE, g_i Foreground SE, b_i

The cells of the structuring element without negation are placed in the foreground set. The cells with negation are inverted and placed in the background set. These are applied to the background of the image.

The structuring elements without negation, i.e., those corresponding to increasing functions, simply have an empty background set. Therefore, in morphological and set notation the operation is written as

$$\bigcup_i \left[(I \Theta b_i) \bigcap (I' \Theta g_i) \right], \tag{5.9}$$

where b_i and g_i are the corresponding structuring element pairs in the foreground and background sets respectively, Θ is the morphological erosion and I' is the inverted image such that the structuring element g_i is applied to the background pixels.

Increasing filters tend to work well in removing certain types of signal-independent additive noise. Bear in mind that the more black pixels there are in the input window, the greater the possibility there is that the output pixel should be black. If the observation tables shown in earlier chapters were mapped onto the lattice of the function, they reflect this trend, otherwise an increasing filter is of no use. Increasing filters are no good for recognition-type problems. Consider the earlier OCR example that attempted to find the letter "e". For an all-white input window the output should be 0. Similarly, for an all-black window it should also be 0. It is only for some particular cases of input fitting the letter "e" that the output should be 1. So an increasing filter would not work in this case and a nonincreasing would be necessary. Dougherty[8] showed that any nonincreasing filter may be expressed as the difference between two increasing filters. This is similar to the hit-or-miss transform where one filter characterizes the foreground and one the background. The two filters must however be designed together and not separately.

Returning to the earlier example of image restoration shown in Fig. 2.1, the optimum filter for this image was determined from the observation table shown in Fig. 3.8. Using minimization techniques, it can be shown that the optimum function reduces to the expression shown in Eqn. 5.10:

$$F = X_0X_1X_2 + X_0X_2X_3 + X_1X_2X_4 + X_2X_3X_4. \tag{5.10}$$

$$F = X_2(X_0X_1 + X_0X_3 + X_1X_4 + X_3X_4).$$

As the function has no negation, it is an increasing function and therefore has a morphological basis representation. The structuring elements to implement this morphological representation are shown in Fig. 5.6. These structuring elements give a great insight into the nature of filtering being applied. In all of the structuring elements, the center pixel X_2 is black. Therefore, only pixels that are black prior to filtering will be black after filtering. Effectively, it will switch some black pixels to white but not the other way around. This makes sense because it was trained just on additive noise and will only try to remove it. The structure is also very interesting. For a black pixel to be retained, it must be supported by two other pixels. However, these two cannot be opposite each other.

Having placed the filter in a morphological context, it is a simple matter to implement an electronic circuit to carry out the filtering. The union of erosions translates directly into a sum-of-products implementation of the filter. One four-input OR gate fed by four three-input AND gates completes the circuit. This is shown in Fig. 5.7. Notice that there are no inverters anywhere in the circuit. This is largely a schematic representation, but it can very easily be converted to discrete hardware, FPGA, or ASIC implementation.

In restoration examples like this, the number of corrupted pixels is usually a small proportion of the total document—typically 5–15%. If the error rate was much greater than this (say 50–60%), statistical restoration would be of no value since the noise would be in the majority. Therefore, only a minority of pixels are

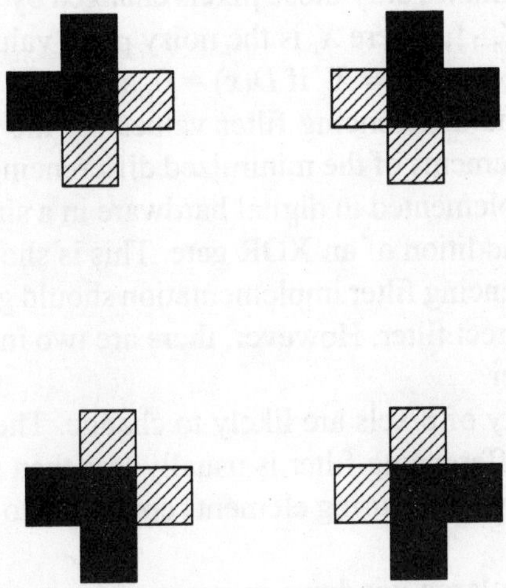

Figure 5.6 Structuring elements to implement the optimum filter for the additive noise example from Fig. 2.1.

likely to be changed in any given document. Bearing this in mind, an alternative implementation is possible known as a *differencing* filter, $D(\mathbf{x})$. This is defined as

$$\psi(\mathbf{x}) = I \otimes D(\mathbf{x}), \tag{5.11}$$

where $\otimes$ is the exclusive OR operator.

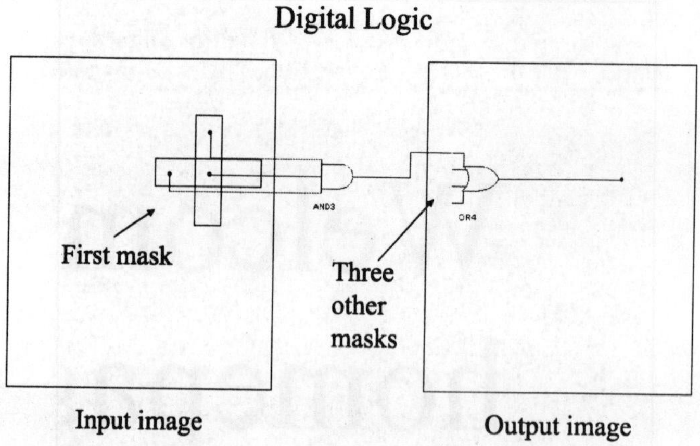

Figure 5.7 Digital logic implementation of the additive noise filter for the example in Fig. 2.1.

Whereas the filter ψ is designed to estimate the pixel value in the ideal image, the differencing filter D estimates only those pixels changed by filtering. For example, if $\mathbf{x} = \{X_0, X_1 \ldots, X_c, ..X_{n-1}\}$, where X_c is the noisy pixel value at the center of window, then the filter output $\psi(x) = X_c$ if $D(x) = 0$, and $\psi(x) = \overline{X}_c$ if $D(x) = 1$.

Figure 5.8 shows the differencing filter values for the previous filter. It also shows the structuring elements of the minimized differencing function. The differencing filter may be implemented in digital hardware in a similar way to the direct filter but including the addition of an XOR gate. This is shown in Fig. 5.9.

In theory, the differencing filter implementation should give precisely the same filtering results as the direct filter. However, there are two main reasons for choosing the differencing filter.

First, only a minority of pixels are likely to change. Therefore, the amount of logic required for the differencing filter is usually less than for the direct filter. In this case, there are just two structuring elements compared to four for the direct implementation.

Second, when using large windows in practice (as seen in Chapter 4), there may be some input combinations that have not been seen during training. In these cases, it is not clear which value to allocate to the output. In the direct filter implementation, these unseen values may be given an arbitrary value resulting in, on average, 50% error for these inputs. A strategy that appears to give improved results in practice involves leaving the input pixel unchanged. The value of pixels is only changed when there is strong statistical evidence to do so. Using the differencing filter design, these unseen inputs are allocated a value of 0 and so are left unchanged by filtering.

	X_0	X_1	X_2	X_3	X_4	OPTIMUM	Difference
Minterm	BIT 0	1	2	3	4	FUNCTION	Function
0	0	0	0	0	0	0	0
1	0	0	0	0	1	0	0
2	0	0	0	1	0	0	0
3	0	0	0	1	1	0	0
4	0	0	1	0	0	0	1
5	0	0	1	0	1	0	1
6	0	0	1	1	0	0	1
7	0	0	1	1	1	1	0
8	0	1	0	0	0	0	0
9	0	1	0	0	1	0	0
10	0	1	0	1	0	0	0
11	0	1	0	1	1	0	0
12	0	1	1	0	0	0	1
13	0	1	1	0	1	1	0
14	0	1	1	1	0	0	1
15	0	1	1	1	1	1	0
16	1	0	0	0	0	0	0
17	1	0	0	0	1	0	0
18	1	0	0	1	0	0	0
19	1	0	0	1	1	0	0
20	1	0	1	0	0	0	1
21	1	0	1	0	1	0	1
22	1	0	1	1	0	1	0
23	1	0	1	1	1	1	0
24	1	1	0	0	0	0	0
25	1	1	0	0	1	0	0
26	1	1	0	1	0	0	0
27	1	1	0	1	1	0	0
28	1	1	1	0	0	1	0
29	1	1	1	0	1	1	0
30	1	1	1	1	0	1	0
31	1	1	1	1	1	1	0

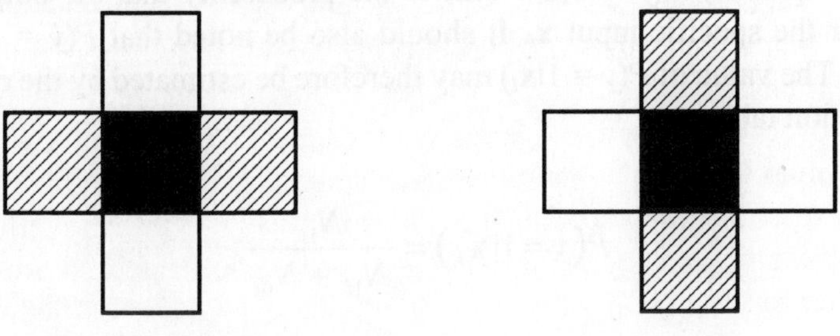

Figure 5.8 Differencing filter design and structuring elements.

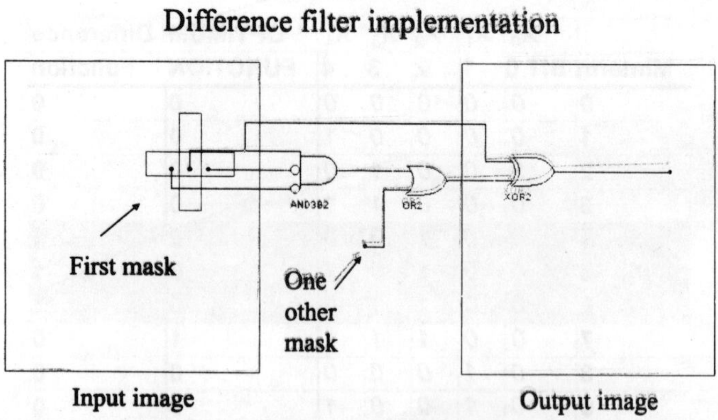

Figure 5.9 Differencing filter implementation.

5.2 Statistical Relevance

The design methods for morphological and logical filters presented in this chapter should not be seen as new or ad hoc approach to filter design. They are, in fact, rooted in standard classical statistics. Consistent with the practical nature of this text, the explanation in this context has been delayed until after the methods have been described by representative examples. In designing the original filters, the process outlined in Chapter 2 involved the compilation of a table of observations which were then used to determine the optimum filter output $\psi_{\text{opt}}(\mathbf{x})$ value for each combination of input values $\mathbf{x}_i$. This is a variation of the conditional expectation filter design method.[6] The method is greatly simplified in this case since both the output and input values are binary. For simplicity, let $y = \psi_{\text{opt}}(\mathbf{x})$.

In order to design the filter, it was necessary to estimate the conditional expectation of the output. This was carried out using the training set. The value of the corresponding pixel in the ideal image was recorded for every input combination. Since the value of y is binary, the conditional output value may be summarized in terms of the quantity $P(y = 1|\mathbf{x}_i)$. This is the probability that the output value y equals 1 for the specific input $\mathbf{x}_i$. It should also be noted that $P(y = 0|\mathbf{x}_i) = 1 - P(y = 1|\mathbf{x}_i)$. The value of $P(y = 1|\mathbf{x}_i)$ may therefore be estimated by the counts from the observation table as

$$\hat{P}(y = 1|\mathbf{x}_i) = \frac{N_{1i}}{N_{1i} + N_{0i}}, \tag{5.12}$$

where N_{1i} and N_{0i} are the counts for $y = 1$ and $y = 0$, respectively, for a specific line i in the observation table.

Num	$P(x_i)$	X_0	X_1	X_2	P(y = 0)	P(y = 1)
0	0.997380	0	0	0	0.999953	0.000047
1	0.000295	0	0	1	0.947368	0.052632
2	0.000171	0	1	0	0.909091	0.090909
3	0.000202	0	1	1	0.153846	0.846154
4	0.000295	1	0	0	0.842105	0.157895
5	0.000078	1	0	1	0.200000	0.800000
6	0.000202	1	1	0	0.076923	0.923077
7	0.001380	1	1	1	0.022472	0.977528

Figure 5.10 Table of observations from Fig. 2.8 expressed in terms of statistics.

The probability that any particular input will occur is $P(\mathbf{x}_i)$. The prior probability may also be estimated as

$$\hat{P}(\mathbf{x}_i) = \frac{N_{1i} + N_{0i}}{\sum_i N_{1i} + N_{0i}}. \tag{5.13}$$

The values of the observation table given in Fig. 2.8 have been reorganized into probabilities in Fig. 5.10.

In order to minimize the MAE, the output of the filter must be the one which is correct most often. Therefore,

$$\begin{aligned} \psi(x_i) &= 1 \text{ if } P(y_i = 1 \mid \mathbf{x}_i) \geq 0.5 \text{ and} \\ \psi(x_i) &= 0 \text{ if } P(y_i = 1 \mid \mathbf{x}_i) < 0.5. \end{aligned} \tag{5.14}$$

This is directly equivalent to selecting the output corresponding to the larger of the two observation values N_{0i} and N_{1i} for each input i. Therefore, the method described in Chapter 2 corresponds to a practical implementation of the maximum likelihood approach.

5.3 Summary

This chapter has introduced the idea of filter function constraints. In particular, it has considered increasing functions and presented some of their properties. It has shown that the area of mathematical morphology may be put in the context of increasing filters. More importantly, it has provided a methodology by which the structuring elements of a morphological filter may be designed to implement the optimum filter. It has also shown that nonincreasing filters may be computed through morphology in a way that is equivalent to the hit-or-miss transform. These filters can have either a direct or differencing filter form, and examples of their implementation in digital logic have been given. Finally, a justification of the approaches presented in terms of classical statistics has been presented.

References

1 J. Serra, *Image Analysis and Mathematical Morphology*, Academic Press, New London (1982).
2 J. Serra, *Image Analysis and Mathematical Morphology*, vol. 2, Academic Press, New York (1988).
3 H. J. Heijmans, *Morphological Operators*, Academic Press, New York (1994).
4 G. Matheron, *Random Sets and Integral Geometry*, Wiley, New York (1975).
5 P. Soille, *Morphological Image Analysis*, 2nd ed., Springer, New York (2003).
6 E. R. Dougherty and J. Barrera, "Logical image operators," in *Nonlinear Filters for Image Processing*, E. Dougherty and J. Astola (eds.), 1–60, SPIE Press, Bellingham, WA (1999).
7 M. Sonka, V. Hlavac, R. Boyle, *Image Processing, Analysis and Machine Vision*, London, Chapman Hall (1993).
8 E. R. Dougherty, "Translation-invariant set operators," in *Nonlinear Filters for Image Processing*, E. R. Dougherty and J. Astola (eds.), 99–120, SPIE Press, Bellingham, WA (1999).

Chapter 6

The Median Filter and Its Variants

6.1 The Grayscale Median as a Special Case of a Generalized WOS Filter

The median filter is a much used and sometimes misunderstood tool available to image processing specialists. It should now be clear to readers that the median is not an alternative filter to those described in this text. It is simply a special case, one of many options that might arise from the design techniques should it happen to be the optimum for that example.

As is well known, the standard median filter[1] is formed by rank ordering the samples within the window and selecting the center value. It is a specific case of a generalized weighted order statistic (WOS) filter which may be written as

$$\psi(\mathbf{x}) = T^{\text{th}} \text{ largest } \{W_0 \lozenge X_0, W_1 \lozenge X_1, \ldots\ldots\ldots\ldots\ldots W_{n-1} \lozenge X_{n-1}\} \quad (6.1)$$

where
$W \lozenge X$ means the sample value X repeated W times,
X_i are the input signal sample values associated with each location in the window,
$\mathbf{x}$ is a vector containing the signal samples $\{X_0, X_1 \ldots. X_{n-1}\}$,
W_i are the corresponding filter weights and
T is a threshold value between 0 and $n - 1$.

The general filter described above is a *rank selection filter*[2] in that the output value of the filter always corresponds to one of the inputs. The filter is unable to average or interpolate and this means that it does not produce simple blurring. However, filters with larger windows can give "streaking" effects in images. This will be addressed later in the chapter.

The median filter is good at preserving sharp changes in intensity such as edges. Its rank order properties mean that for impulsive noise, the corrupted pixel values go to the extremes of the distribution and have little or no chance of emerging at the output of the filter. Hence it has strong noise rejection properties. The standard median filter is formed by setting $T = (1 + \sum W_i)/2$ and $W_i = 1$ for all i. It has the property of being self-dual. This means that it treats black and white pixels equally. If the image were to be inverted, then median-filtered and inverted again, it would give the same result as median-filtering the original image.

Two obvious ways of varying the median filter is to change the weights W_i or the threshold value T.

Changing the weights gives more importance to certain pixels—usually those closest to the center of the window. This is important especially for larger windows. A WOS filter that has different values of weights but retains $T = (1 + \sum W_i)/2$ is known as a *weighted median* filter. If the weights are symmetrical about the middle, i.e., $W_i = W_{n-1-i}$, it will also be self-dual.

Changing the threshold parameter T means that a different rank other than the median is chosen. For values of T other than the center value, the filter is not self-dual. If T is allowed to vary, but all the weights are set to $1 (W_i = 1)$, then the filter becomes a rank-order filter. Trivial examples are for $T = 1$ resulting in the minimum and $T = n$ giving the maximum.

In designing WOS filters, the critical question is: What combination of values of W_i and T result in the optimum filter for any given task? It is of course possible to search all values, but this is very time consuming. It is also possible to employ iterative techniques to adjust the filter parameters until the error criterion is minimized. This is necessary in more complicated examples, but in many cases the techniques described in previous chapters may be adapted to determine the optimum parameters. Other work in this field includes Shmulevich[3,4] and Arce[5], which includes filters with negative weights.

All filters that can be put into the context of generalized WOS filters are increasing filters. This means that they have two special properties:

- They may be implemented in terms of mathematical morphology.
- They may be extended to grayscale via threshold decomposition.

The first property may or may not be a useful one. Where the filter results in a simple set of morphological structuring elements, it may be implemented in hardware in terms of comparators, resulting in a fast simple circuit. However, some filters can result in large sets of structuring elements, and more arithmetic-based implementations may give greater efficiency.

The mention of grayscale processing will come as a relief to many readers who feared that they were reading a book limited to binary image processing. Far from it—the techniques will be extended to grayscale in coming chapters. The advantage of several of these techniques is that the optimum grayscale filter may be determined at a binary level through threshold decomposition and then extended to grayscale without loss of optimality. This is a strong property because it drastically simplifies the training.

6.2 Binary WOS Filters

Document images have two distinct gray level distributions corresponding to text and background. After rank ordering, the pixel values corresponding to text go to one end of the list and background pixels to the other. Because the median is the sample at the center of the list, it corresponds to whichever value occurs most often. When applied to binary values it is in effect the majority filter. This causes fine detail in images to be removed. While the median is very good for preserving steps in intensity values, it is poor at preserving the location of such steps.

As explained above, the task now becomes that of designing the optimum binary filter ψ. The objective is still to find the weights W_i and selection parameter T, neither of which is necessarily binary. The behavior of the binary filter ψ may be expressed in terms of the linear summation and inequality shown in Eqn. 6.2.

$$\psi(\mathbf{x}) = \begin{cases} 1, & if \ \sum_{i=0}^{n-1} W_i X_i \geq T \\ 0, & \text{otherwise} \end{cases} \tag{6.2}$$

In the case of binary filters, the standard median is a majority function. This may be expressed in terms of logic. For example, for five variables, three of these are required to be 1, i.e.,

$$\begin{aligned} \psi_{\text{med}} &= X_0X_1X_2 + X_0X_1X_3 + X_0X_1X_4 + X_0X_2X_3 + X_0X_2X_4 \\ &\quad + X_0X_3X_4 + X_1X_2X_3 + X_1X_2X_4 + X_1X_3X_4 + X_2X_3X_4 \, . \end{aligned} \tag{6.3}$$

It can be seen that even for five variables the binary expression is already becoming large. It may therefore be computed as a counting operation rather than a sorting operation. This can result in large reductions in the processing time. It will be shown later that this principle may be extended to grayscale images where the computational advantages are even greater.

6.3 Positive and Negative Medians

Two variations on the median are the *positive* and *negative* median. These filters behave in a similar way to the standard median filter but only allow changes in one direction either from **0** to **1** or **1** to **0**. They are an asymmetric version of the median filter and remove either positive or negative impulses, respectively.

The positive median will retain the center value should it be 1. This may be written as

$$\psi_{+\,\mathrm{med}} = X_2 + \psi_{\mathrm{med}}\,,$$

$$\psi_{+\,\mathrm{med}} = X_2 + X_0X_1X_3 + X_0X_1X_4 + X_0X_3X_4 + X_1X_3X_4\,. \qquad (6.4)$$

For the output pixel to be black, either the center pixel must be black or at least three of the four other pixels must be black.

Alternatively, the negative median will preserve the value of the center pixel should it be 0, otherwise it follows the median and can be written as:

$$\psi_{-\,\mathrm{med}} = \psi_{\mathrm{med}}\,.\,X_2\,.$$

$$\psi_{-\,\mathrm{med}} = X_0X_1X_2 + X_0X_2X_3 + X_0X_2X_4 + X_1X_2X_3 + X_1X_2X_4 + X_2X_3X_4\,. \qquad (6.5)$$

$$\psi_{-\,\mathrm{med}} = X_2(X_0X_1 + X_0X_3 + X_0X_4 + X_1X_3 + X_1X_4 + X_3X_4)\,.$$

For the negative median filter, the center pixel will only be black if it is black prior to filtering and supported by at least two other black pixels.

6.4 Weighted Median Filters

Median filters give equal weighting to all of the pixels within the window. As mentioned earlier, this can cause streaking effects especially for larger window sizes. This effect can be reduced by giving more importance to the pixels close to the center of the filter window.

It is often claimed that the median filter is good at preserving edges. This is only half true. In grayscale images that would otherwise be blurred by Gaussian or other linear smoothing filters, the abrupt height of the step of the edge is preserved. However, the position of the edge can be shifted to a different location.

Figure 6.1 shows two examples of image detail that may be damaged by median filtering. The first shows a corner pixel removed from a 90 degree angle. This may be preserved using a weighted median. The second effect is known as edge “pulling”. Isolated noise pixels close to an edge can cause it to be pulled out at this point.

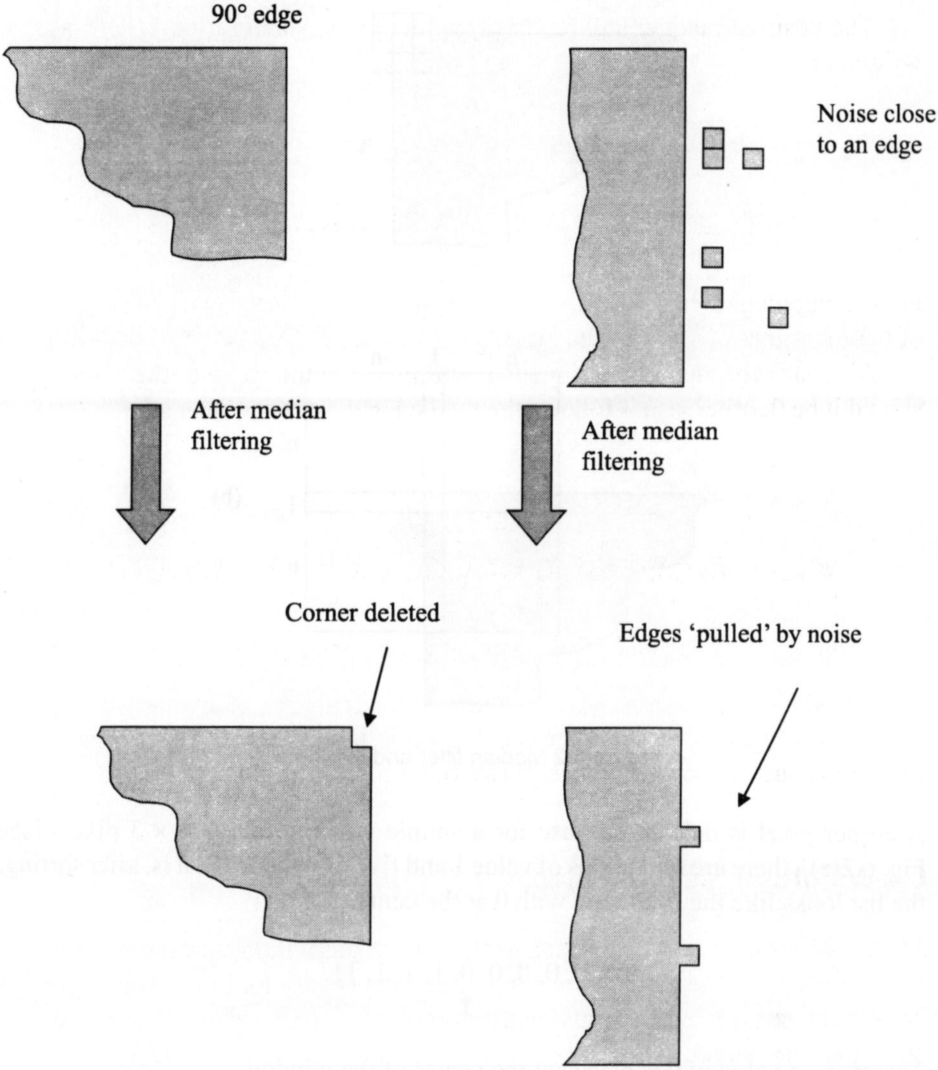

Figure 6.1 Median filter and fine detail. Two examples of image detail that may be damaged by median filtering. The first shows that a corner pixel will be removed from a 90-degree angle. This may be preserved using a weighted median filter. The second effect is known as edge "pulling". Isolated noise pixels close to an edge can cause it to be pulled out at this point.

6.5 Optimum Design of Weighted Rank and Median Filters

Where certain fine structures must be preserved within an image, the filter weighting can be chosen to do this. For example, consider the case of corner preservation in median filtering. Consider the foreground pixel to have value 1 and the background 0.

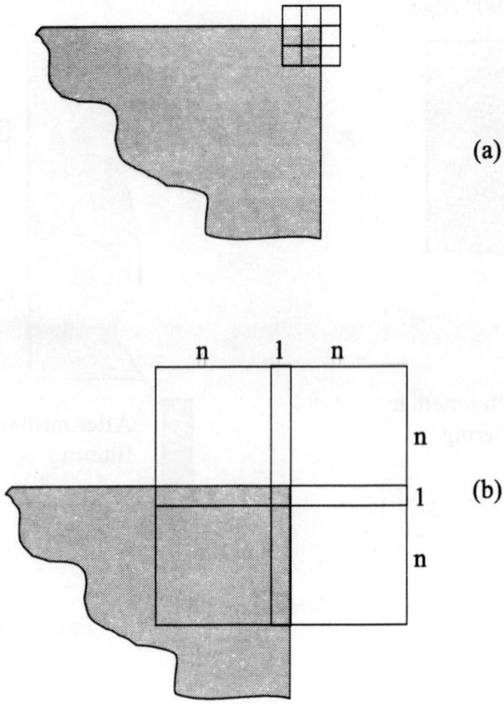

Figure 6.2 Median filter and fine detail.

A corner pixel is deleted because for a simple window of say 3×3 pixels [see Fig. 6.2(a)], there are four pixels of value 1 and five of value 0. That is, after sorting, the list looks like the following with 0 at the center:

$$\{0, 0, 0, 0, \underset{\uparrow}{0}, 1, 1, 1, 1\}.$$

Therefore, a value of 0 is placed at the center of the window.

This effect may be removed by giving the center pixel a weighting of 3. Hence, it is placed in the list three times and results in the following list:

$$\{0, 0, 0, 0, 0, \underset{\uparrow}{1}, 1, 1, 1, 1, 1\}$$

with the output now 1. The filter thus gives more importance to the center pixel. However this supresses the noise reduction properties. The center weighting may be calculated for larger windows as shown in Fig. 6.2(b). For an odd-sized square window of dimension $(2n + 1) \times (2n + 1)$ placed at the corner pixel, the number of background pixel values in the list corresponds to

$$3n^2 + 2n\,. \tag{6.6}$$

Furthermore, the number of foreground pixels placed into the list is

$$n^2 + 2n + W, \tag{6.7}$$

where W is the weighting given to the center pixel. In order that the corner pixel is preserved

$$n^2 + 2n + W > 3n^2 + 2n \tag{6.8}$$

Since all quantities are integers, the critical point occurs when

$$n^2 + 2n + W = 3n^2 + 2n + 1, \tag{6.9}$$

therefore $$W = 2n^2 + 1.$$

This is consistent with a weighting of 3 for a 3×3 window. Other weightings are given in Table 6.1.

Weightings may be calculated for the preservation of other fine detail in a similar way. In general, for larger windows weightings are to be applied to other window locations in addition to the center. The weightings may be determined by forming and solving a series of simultaneous equations. This may also be carried out for more general weighted rank-order filters where the rank is also a parameter.

The method described in Chapter 2 may be extended for the design of optimum WOS filters and weighted median filters. In Chapters 2 and 3, a table of observations was generated and the optimum filter was derived. The filters were unconstrained in their function and the output was set independently for every input combination. WOS and median filters represent a constraint on the function that may be implemented by restricting the output to depend on a weighted summation

Table 6.1 Weighting required for corner preservation for various sizes of center-weighted median filters.

Window size		**Weighting**
$(2n+1)\times(2n+1)$	n	$2n^2+1$
3×3	1	3
5×5	2	9
7×7	3	19
9×9	4	33

and thresholding operation. The MAE of a WOS or median filter will therefore be either greater than or the same as an unconstrained filter implemented within the same-sized window.

The unconstrained filter implemented in a window of n pixels has 2^n independent input combinations. The WOS and median filters have a much smaller set of possible inputs. Taking the special case of a simple rank-order filter $\psi(\mathbf{x})$, all inputs with the same Hamming weight are in effect the same input. The Hamming weight[6] is the sum of the pixel values in the filter window, i.e., $|\mathbf{x}| = \Sigma_i X_i$. Therefore, for a 5-input window, the inputs $\mathbf{x} = (0,0,0,0,1)$, $\mathbf{x} = (0,0,0,1,0)$, $\mathbf{x} = (0,0,1,0,0)$, $\mathbf{x} = (0,1,0,0,0)$, $\mathbf{x} = (1,0,0,0,0)$ all result in the same output. This means that many inputs for the unconstrained filter map to a single input in the rank-order filter. The filter may therefore be written as a function of the Hamming weight of the input vector, $\psi(|\mathbf{x}|)$. There are in effect just $n + 1$ inputs rather than 2^n. The observation table may therefore be written with just $n + 1$ lines.

Consider the images of Fig. 6.3. A 3×3 WOS filter results in an observation table with just 10 ($|\mathbf{x}| = 0........9$) lines compared to $512(= 2^9)$ lines for the unconstrained function. In the binary case, the simple rank-order filter is equivalent to placing a threshold value r on the Hamming weight of the input $|\mathbf{x}|$ such that the filter output $\psi(|\mathbf{x}|) = 1$ if $|\mathbf{x}| \geq r$, and $\psi(|\mathbf{x}|) = \mathbf{0}$ if $|\mathbf{x}| < r$. The design of the optimum rank filter therefore reduces to the selection of the value r.

Following the same procedure as in previous chapters, the value of r should be set to make the output correspond to the correct value as often as possible. For simplicity, the filter output value will be written as y. It can be seen from the table of observations in Fig. 6.3(a) and the corresponding probabilities shown in Fig. 6.3(b) that the probability of the filter output being $\mathbf{1}$, $p(y = 1|\,|\mathbf{x}|)$ is seen to increase monotonically with $|\mathbf{x}|$. The value of r_{opt} that results in the optimum rank filter $\psi(|\mathbf{x}|)_{\text{opt}}$ therefore corresponds to the minimum value of $|\mathbf{x}|$ for which $p(y = 1|\,|\mathbf{x}|) \geq 0.5$. In this case $r_{\text{opt}} = 6$. Selecting any other rank, including the median ($r = 5$) will result in a filter with an increased error compared to $\psi(|\mathbf{x}|)_{\text{opt}}$ ($r = 6$). The number of pixels in error in the image filtered by the optimum rank filter may be found by summing the minimum value of N_0 and N_1 from each line of the table of observations. These are shown in Fig. 6.3(a) and the appropriate value is shaded in gray. A comparison of the noisy image filtered by the optimum filter and by the median filter is shown in Fig. 6.3(c) and 6.3(d). The median has an additional 24 pixels in error corresponding to the difference between N_0 and N_1 (308 and 284) in line $|\mathbf{x}| = 5$ in the observation table.

6.6 Weight-Monotonic Property

It was seen in Fig. 6.3(b) that the probability of the filter output being $\mathbf{1}$, $p(y = 1|\,|\mathbf{x}|)$ increased monotonically with $|\mathbf{x}|$, i.e.,

(a)

$\|x\|$	$N(y = 0)$	$N(y = 1)$
0	22610	0
1	22607	0
2	10242	17
3	3397	30
4	1097	48
5	308	284
6	60	995
7	5	629
8	0	461
9	0	1726

(b)

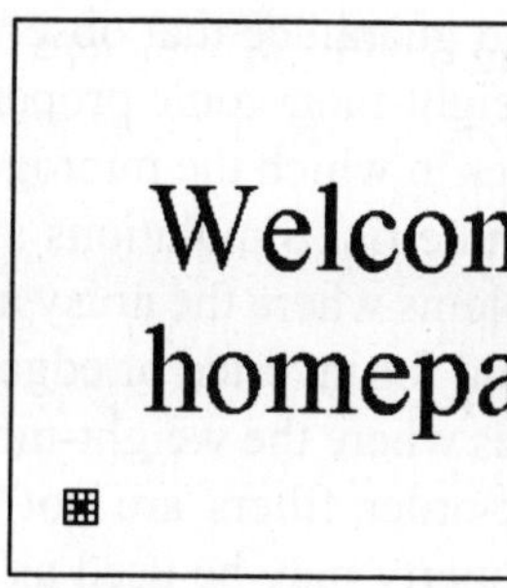

$\|x\|$	$P(\|x\|)$	$P(y = 0\| \|x\|)$	$P(y = 1\| \|x\|)$	
0	0.35	1.00	0.00	0
1	0.35	1.00	0.00	0
2	0.16	1.00	0.00	0
3	0.05	0.99	0.01	0
4	0.02	0.96	0.04	0
5	0.01	0.52	0.48	0
6	0.02	0.06	0.94	1
7	0.01	0.01	0.99	1
8	0.01	0.00	1.00	1
9	0.03	0.00	1.00	1

(c)

Error = 468 pixels

(d)

Error = 444 pixels

Figure 6.3 (a) shows noisy and ideal image observations and (b) gives the probability estimates that show that the optimum rank-order filter occurs when $r_{opt} = 6$. The noisy image filtered by the median filter ($r = 5$) is shown in (c) and the noisy image filtered by the optimum filter ($r = 6$) is shown in (d).

$$P(y = 1 \,|\, |\mathbf{x}| = i) \geq P(y = 1 \,|\, |\mathbf{x}| = j) \quad for \quad i > j \tag{6.10}$$

This property is known as the *weight-monotonic* property[7] and implies that the more black pixels there are in the observation window, the more likely it is that the ideal pixel at the window center is black.

There is no guarantee that observations collected from any given test set will possess the weight-monotonic property. The model, however, is not unreasonable for ideal images in which the microgeometry is somewhat random and the noise is white and symmetric. Simulations show that these assumptions hold for restoration-type problems where the noisy and ideal images have similar pixel values, but they do not hold for inverted or edge-detected images.

In the cases where the weight-monotonic property does not hold, it would suggest that rank-order filters are not applicable for these problems. The weight-monotonic property may be used as a test to check if increasing filters in general and rank-order filters in particular are suitable for a given problem.

6.7 Design of Weighted Median Filters

The previous example showed the design of a filter constrained to be a rank-order filter. This approach may be extended to weighted median filters.

Weighted median filters are self-dual. This means that they treat black and white pixels equally. Therefore, as well as constraining the filter to depend on a weighted ordering of its inputs, it must also be constrained to be self-dual. These constraints may be enforced by placing the design problem in the context of a differencing filter, *D*. The design of the optimum center-weighted median filter within a window *B* reduces to the problem of determining the pixel weighting *W* for which the *MAE* is a minimum.

As a result of the constraint of *self duality*, it is easier to analyze the weighted median filter by considering the conditions under which the center pixel, X_c switches state, either from **0** to **1** or vice versa. This is done by defining W_{med} in terms of a differencing filter $D(\mathbf{x})$:

$$W_{\text{med}} = X_c \otimes D(\mathbf{x}), \tag{6.11}$$

where $\otimes$ is the set difference (XOR) operator.

Rather than specifying an absolute value of 0 or 1, the differencing filtering $D(\mathbf{x})$ indicates whether the value at the center of the window X_c should be changed. Examples of the four cases of $D(\mathbf{x})$, X_c, and W_{med} are given in Table 6.2.

Table 6.2 Operation of the differencing filter $D(\mathbf{x})$.

Pixel at Center of window, X_c	Differencing filter value, $D(\mathbf{x})$	Output of Weighted Median Filter, W_{med}
0	0	0
0	1	1
1	0	0
1	1	1

The differencing filter is sometimes described as a toggle filter and any translation invariant filter may be put into this context. The differencing filter $D(\mathbf{x})$ is therefore equal to 1 at those locations where the pixel value at the center of the window X_c is changed by filtering and 0 where it is unchanged. The process of designing the optimum differencing filter reduces to the task of determining if the resulting MAE will be lower by switching the value of X_c or leaving it unchanged.

Consider a 3×3 filter window. The pixel X_c at the center of the window has eight neighbors. In noise reduction problems such as those addressed by the weighted median filter, the value of the pixel X_c is considered to be noise and switched if a sufficient number of its neighboring pixels have the opposite value. For example, if $X_c = 0$ and most of its surrounding pixels have value 1, then there is a strong case that its value should be changed to $X_c = 1$. As the WM filter is increasing, it will cause a pixel with value $X_c = 0$, to switch to $X_c = 1$ if the number of surrounding pixels with value 1 exceeds a given threshold. The value of the threshold is directly related to the filter weight.

The important question is: Which is the optimum filter weight? That is, which weight results in the filter giving the lowest MAE? This may be determined from a representative training set similar to the approach taken for the WOS filter. The effect on the resulting MAE of varying the filter weight may be evaluated.

This is carried out as follows: Let $|\mathbf{x}'|$ be the number of pixels in the filter window having the opposite value to the center pixel X_c. To clarify, some examples are given in Fig. 6.4. In Fig. 6.4(a), the value of $X_c = 0$. This is because the center pixel has value 0, and $|\mathbf{x}'| = 2$ as there are 2 pixels of value 1 (the opposite of X_c). Similar results are obtained in Figs. 6.4 (b) and (c).

For a standard median filter implemented in a 3×3 window, at least five neighboring pixels are required to have the opposite value to the center pixel, i.e., $|\mathbf{x}'| = 5$ to cause it to switch state. Since the WMF is self-dual, the conditions for it to switch in either direction are the same. For each increase in center weight, one further

0	0	0
1	0	0
0	0	1

(a)

$X_c = 0$

$|\mathbf{x}'| = 2$

1	1	1
0	1	1
0	1	0

(b)

$X_c = 1$

$|\mathbf{x}'| = 5$

0	0	0
0	1	1
0	0	0

(c)

$X_c = 1$

$|\mathbf{x}'| = 7$

Figure 6.4 Examples of various values of X_c and $|\mathbf{x}'|$. In (a) $X_c = 0$. This is because the center pixel is 0, and $|\mathbf{x}'| = 2$ since there are 2 pixels of value 1 (the opposite of X_c). Similarly for (b) $X_c = 1$, $|\mathbf{x}'| = 5$, and (c) $X_c = 1$, $|\mathbf{x}'| = 7$. The higher the value of $|\mathbf{x}'|$, the greater the probability that X_c will switch value during filtering.

Table 6.3 Switching strength of weighted median filter for various center weightings.

Center weight, W	Number of neighbors of opposite state required to cause center pixel to switch state
1	5
3	6
5	7
7	8
>7	Not possible

neighboring pixel is required to trigger a switch. It can easily be shown that there are therefore only four valid center weights for the filter defined in a 3×3 window and that these are 1, 3, 5, and 7. When the center weighting is $W = 1$, the filter is identical to the standard median. For center weights greater than seven, it becomes impossible to switch the center value even if all other eight pixels have the opposite value. In this case, the filter becomes an identity filter and it is neither extensive nor antiextensive. This relationship is shown in Table 6.3.

For simplicity, let $d = D(\mathbf{x})$ and $P(d = 1||\mathbf{x}'|)$ be the probability that X_c will switch value when a total corresponding to $|\mathbf{x}'|$ of its neighbors have the opposite value. Similarly, $P(d = 0||\mathbf{x}'|)$ is the probability that X_c will remain unchanged under the same conditions. The prior probability of $|\mathbf{x}'|$ is given by $P(|\mathbf{x}'|)$ and $P(d = 1||\mathbf{x}'|) = 1 - P(d = 0||\mathbf{x}'|)$.

Assuming that the weight-monotonic property holds, then the probability that a pixel will switch state $P(d = 1||\mathbf{x}'|)$ increases monotonically with the number of neighbors it has of the opposite value $|\mathbf{x}'|$. It is expected that this property would be reflected in the training set data for an imaging problem capable of being corrected by an increasing filter such as the weighted median.

By the same argument as in the general case and the WOS filter, the optimum differencing filter $D_{\text{opt}}(I)$ is determined by $|\mathbf{x}'|_{\text{opt}}$, the minimum value of $|\mathbf{x}'|$ for which $P(d = 1||\mathbf{x}'|) \geq 0.5$. The total MAE may be calculated in a similar way as will be seen.

The probability that the center value will switch $P(y \neq X_c\, ||\mathbf{x}'|)$ is used to design the differencing filter and is estimated from the training set. A variation on the familiar table of observations is formed. The training images are scanned with the filter window and at each location a count is kept as to whether the ideal image value y differs from the noisy value X_c for each value of $|\mathbf{x}'|$ in the window. The switching probability $P(y \neq X_c\, ||\mathbf{x}'|)$, is then determined as

$$P(y \neq X_c \| \mathbf{x}'|) = \frac{N_{y \neq X_c}(|\mathbf{x}'|)}{N_{y \neq X_c}(|\mathbf{x}'|) + N_{y = X_c}(|\mathbf{x}'|)}, \tag{6.12}$$

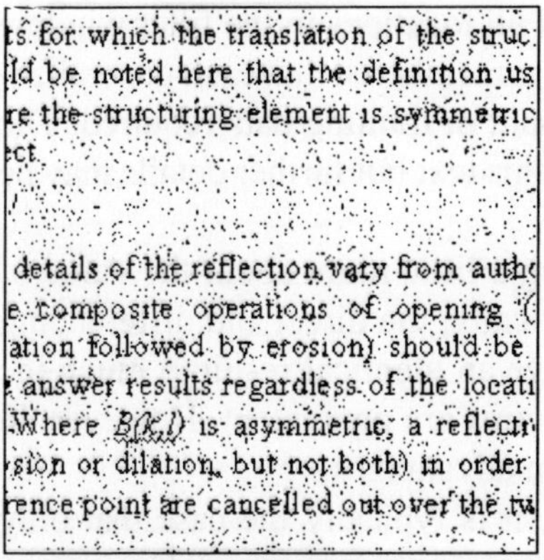

(a) Original noisy image

$\lvert x'\rvert$	$P(\lvert x'\rvert)$	$P(d = 0, \lvert x'\rvert)$	$P(d = 1\mid \lvert x'\rvert)$	$D(x)$
0	0.45	1.00	0.00	0
1	0.22	1.00	0.00	0
2	0.09	1.00	0.00	0
3	0.07	0.99	0.01	0
4	0.05	0.97	0.03	0
5	0.04	0.91	0.09	0
6	0.05	0.89	0.11	0
7	0.01	0.22	0.78	1
8	0.02	0.01	0.99	1

(b) Probability estimates

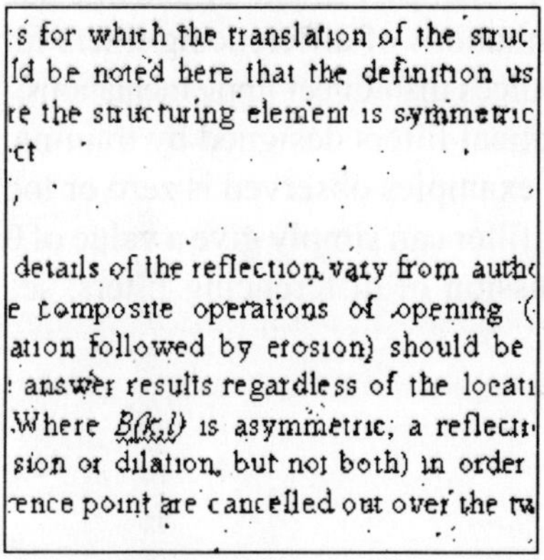

(c) Weighted median filter $W = 5$ (optimum).

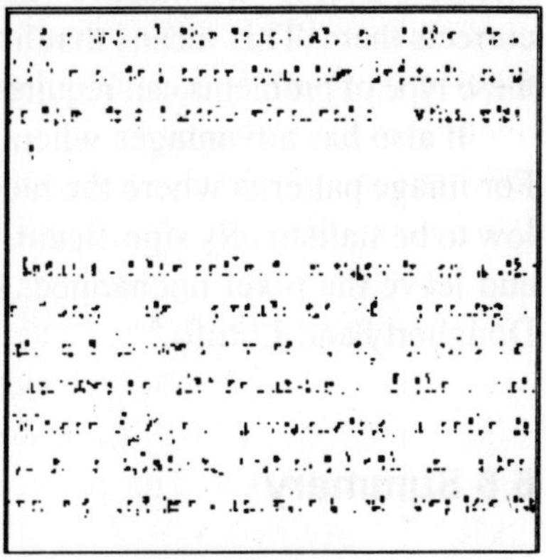

(d) Standard median filter

Figure 6.5 The detail in image (a) consists of very thin text corrupted by noise. The probability estimates are given in (b). Much of the text is preserved using the optimum weighted median filter (c) with $W = 5$ (equivalent to $|\mathbf{x}| = 7$). However, it is almost destroyed by the standard median filter (d).

where $N_{y=X_c}(|\mathbf{x}'|)$ and $N_{y\neq X_c}(|\mathbf{x}'|)$ are the number of times that the center value switches, i.e., $y \neq X_c$ or is unchanged ($y = X_c$) for input $|\mathbf{x}'|$.

Figure 6.5(a) shows an image containing very thin text corrupted by noise. The probability estimates are given in Fig. 6.5(b). These show that a value of $|\mathbf{x}'|_{opt} = 7$ gives the optimum weighted median. This is the minimum value of $|\mathbf{x}'|$ for which $P(d = 1 \,|\, |\mathbf{x}'|) \geq 0.5$ and corresponds to a filter weight of $W = 5$. The result of applying the optimum weighted filter is shown in Fig. 6.5(c), where it can be seen that most of the text is preserved. In contrast, applying the standard median destroys most of the text and results in the image shown in Fig. 6.5(d). The filters with weights on either side of the optimum, i.e., $W = 3$ and 7, were found to give very poor results, suggesting that the selection of the optimum weight is critical in this case.

It is possible to generalize the filter such that all locations in the filter window may be allocated individual weights. It should be remembered that each generalization will bring an associated increase in the amount of training data required. The details of these methods are beyond the scope of this text but further details may be found in Marshall.[7]

Before closing the chapter, it is worth saying a little more about the value of differencing filters in image processing. Theoretically, the direct and differencing representation of a filter are equivalent. They give an identical result in much the same way as a sum of products and product of sums are equivalent. However, in practice the differencing filter can possess certain advantages.

In image restoration problems, for example, it is typically the case that only 10–20% of the pixels are corrupted and therefore require correction. This means 80–90% of the pixels should remain unchanged by filtering. The differencing filter is therefore a relatively inactive filter—it identifies a small percentage of patterns and corrects them. This means that hardware implementations of differencing filters for these type of problems can require much fewer resources than direct implementations.

It also has advantages when extended to practical filters designed by training. For image patterns where the number of training examples observed is zero or too low to be statistically significant, the differencing filter can simply give a value of 0 and leave the pixel unchanged. For further discussion of differencing filters, see Dougherty and Lotufo.[8]

6.8 Summary

This chapter has described two variations on the median filter. They both attempt to make it more flexible either for the rejection of noise or the preservation of image detail. They involve allowing the filter weights and the threshold parameter to vary. Design methods have been presented for these filters based on the weight-monotonic property. The differencing filter has been introduced to ensure that the weighted median filters are self-dual.

Optimum design of both the weighted order statistic (WOS) filters and the weighted median filters (WMF) are not restricted to binary images and may be extended to grayscale[9] processing via the threshold decomposition theorem. This is explained in the next chapter.

References

1 M. Gabbouj, E. Coyle, and N. Gallagher, Jr., "An overview of median and stack filtering," *Circuits, Systems, and Signal Processing*, **11**(1), 7–45 (1992).

2 P. Maragos and R. W. Schafer, "Morphological filters—Part I: Their relations to medians, order statistics, and stack filters," *IEEE Trans. Acoustics, Speech, and Signal Processing*, **35**, 1153–1169 (1987).

3 I. Shmulevich, V. Melnik, and K. Egiazarian, "The use of sample selection probabilities for stack filter design," *Signal Processing Letters*, **7**(7), 189–192 (2000).
4 I. Shmulevich and G. R. Arce, "Spectral design of weighted median filters admitting negative weights," *Signal Processing Letters*, **8**(12), 313–316 (2001).
5 G. R. Arce, "A general weighted median filter structure admitting negative weights," *Signal Processing*, **46**(12), 3195–3205 (1998).
6 R. Hamming, "Error-detecting and error-correcting codes," *Bell System Technical Journal*, **29**(2), 147–160 (1950).
7 S. Marshall, "A new direct design method for weighted order statistic filters," *IEE Proceedings on Vision, Image and Signal Processing*, **151**(1), 1–8 (2004).
8 E. R. Dougherty and R. Lotufo, *Hands-on Morphological Image Processing*, SPIE Press, Bellingham, WA (2003).
9 O. Yli-Harja, J. Astola, and Y. Neuvo, "Analysis of the properties of median and weighted median filters using threshold logic and stack filter representation," *IEEE Trans. Signal Processing*, **39**(2), 395–410 (1991).

Chapter 7

Extension to Grayscale

The chapters of this book have thus far mainly addressed binary image processing. While binary image processing is useful in some circumstances, it is very limited in its applications. The challenge facing nonlinear image processing is to take these methods and extend them to grayscale. This can be done in a number of different ways. Current approaches are listed below:

- Stack filters
- Grayscale morphology
- Computational morphology
- Aperture filters

7.1 Stack Filters

Stack filters were introduced by Wendt, Coyle, and Gallagher[1] at Purdue in the 1980s. They enable the transition between binary and grayscale processing through a concept known as threshold decomposition.[2] In digital systems where a signal is represented in a finite number of bits, a grayscale signal X consisting of m discrete levels may be thresholded at every level to produce $m - 1$ binary signals $\mathbf{x}^t$, i.e.,

$$\mathbf{x}^t = [X]^t$$

where $[\ ]^t$ is the thresholding operator, defined as

$$[\ ]^t = \begin{cases} 1 & if \quad [\ ] \geq t \\ 0 & if \quad [\ ] < t \end{cases} \tag{7.1}$$

An example of threshold decomposition is given in Fig. 7.1.

Note: there is usually one less binary signal $\mathbf{x}^t$ than the number of gray levels in X because thresholding at the bottom level 0 results in the trivial binary signal $\mathbf{x}^0$ for which every value is equal to 1. This is sometimes omitted.

Threshold Decomposition

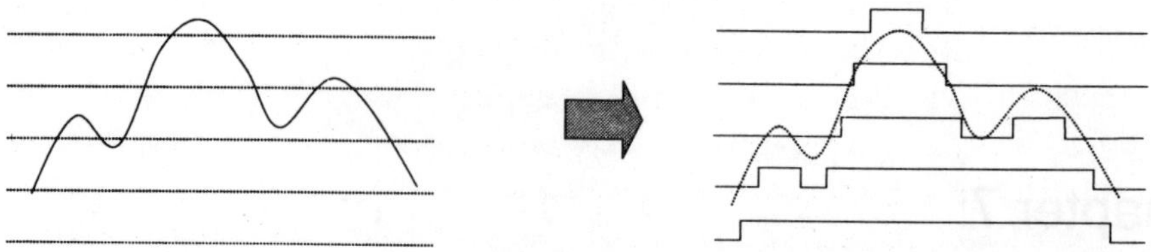

A grayscale waveform of m levels is transformed to $m - 1$ binary signals

Figure 7.1 Threshold decomposition.

The binary signals $\mathbf{x}^t$ are known as a *stack*. They may be processed using a binary filter ψ to produce a series of binary output signals $\mathbf{y}^t$,

$$\psi(\mathbf{x}^t) = \mathbf{y}^t. \tag{7.2}$$

These may be summed (or *stacked*) to give a grayscale output signal Y as follows:

$$Y = \sum_t y^t. \tag{7.3}$$

An example of stacking a set of binary signals to give a grayscale signal is shown in Fig. 7.2. For a certain class of filters, the grayscale output signal Y resulting from the process of thresholding followed by binary filtering and then summation is precisely the same as that which results from filtering the grayscale signal X with the grayscale version of the filter ψ. The class of filters for which this holds includes many types of filters such as WOS, including the median, weighted median, and rank-order filters.

The ability to decompose a grayscale function into a series of binary operations can be a valuable one. As shown in the previous chapter, the binary median filter may be implemented as a counting operation rather than sorting. The stack filter allows this property to be utilized even for grayscale signals. The decomposition may be useful for proving theorems and characterizing filters. It is rarely, however, of direct use in implementation.

An example of a 3-point running median (with three levels) implemented as a grayscale operation and via a stack filter is shown in Fig. 7.3. The grayscale signal is median filtered by two routes. The first is by applying threshold decomposition to process a stack of binary signals. These are individually filtered using a 3-point binary running median. The binary median consists of a simple counting operation. The resulting binary outputs are then stacked to produce the grayscale output signal. This is precisely the equivalent to the output of the 3-point grayscale running median applied to the original signal and implemented via rank ordering. A sorting operation has thus been replaced by a counting operation which may in some cases prove a useful property for more efficient implementation.

Filters that give the same answer through direct grayscale and stack filter implementation are said to obey the stacking property or to commute with thresholding. For a filter to commute with thresholding, its binary version must be based on a *positive Boolean function (PBF)*. That is, the function must be capable of being

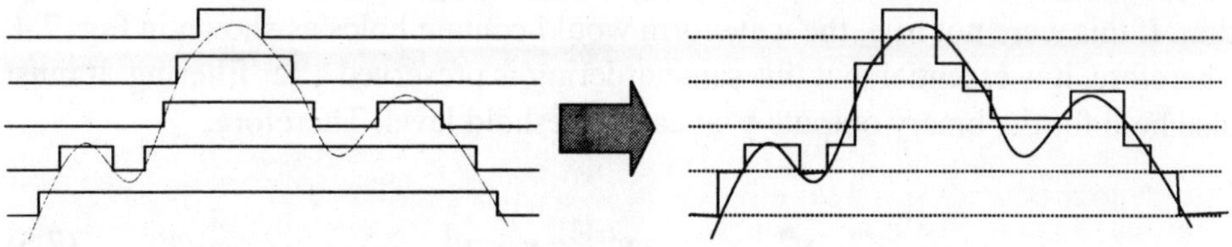

Figure 7.2 Stacking. $m - 1$ binary signals are transformed back to an m-level grayscale waveform.

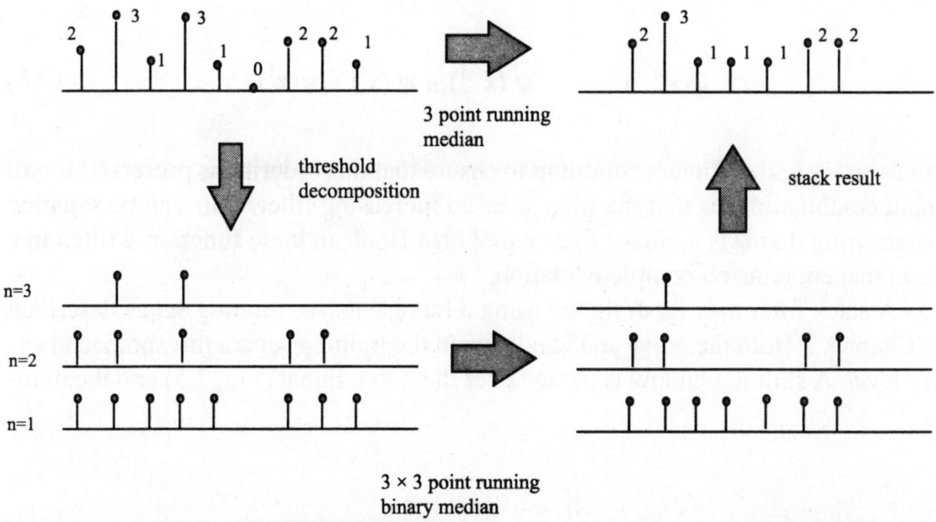

Figure 7.3 Stack filter implementation of a 3-point running median. The stack filter is implemented via two separate routes. Starting at the top left the signal may be transformed into three binary signals via threshold decomposition (bottom left). These are then individually filtered using a 3-point running binary median, which is in effect a counting operation. The resulting binary signals (bottom right) may then be stacked to produce the median filtered grayscale signal (top right). This is precisely the same result achieved by carrying out a 3-point running grayscale median filter, based on rank ordering. The process of threshold decomposition has thus replaced a sorting operation by three separate counting operations.

written in terms of its variables without complementation. This is equivalent to the increasing property as was seen in Chapter 5. The reason for this is straightforward. After threshold decomposition every binary signal in the stack X^t is included within the one beneath., i.e.,

$$\mathbf{x}^{m-1} \leq \ldots\ldots\ \mathbf{x}^{t+1} \leq \mathbf{x}^t \leq \mathbf{x}^1 \tag{7.4}$$

It is in the nature of all real signals that a strict ordering as specified in Eq. 7.4 can be observed within a stack of binary signals resulting from threshold decomposition. If this were not true, the waveform would contain holes as shown in Fig. 7.4. Therefore, it is essential that this same ordering is preserved after filtering. It must also hold for the binary outputs $\mathbf{y}^t$ at each threshold level. Therefore,

$$\mathbf{y}^{m-1} \leq \ldots\ldots\ \mathbf{y}^{t+1} \leq \mathbf{y}^t \leq \mathbf{y}^1. \tag{7.5}$$

This in turn leads to a constraint on the type of filtering that may be applied to the binary levels. Only filters for which the following ordering is preserved may form the basis of a stack filter.

$$\psi(\mathbf{x}^{m-1}) \leq \ldots\ldots\ \psi\,(\mathbf{x}^{t+1}) \leq \psi\,(\mathbf{x}^t) \leq \psi(\mathbf{x}^1). \tag{7.6}$$

A necessary and sufficient condition to ensure that this ordering is preserved for all input combinations is that the filter ψ be an increasing filter. This can be satisfied by ensuring that ψ is a binary filter based on a Boolean logic function written in a form that contains no complementation.

A stack filter may be designed using a representative training set as described in Chapter 2. Both the noisy and ideal data in the training set are thresholded at every level. A sliding window is passed over the noisy signal (Fig. 7.5) and the num-

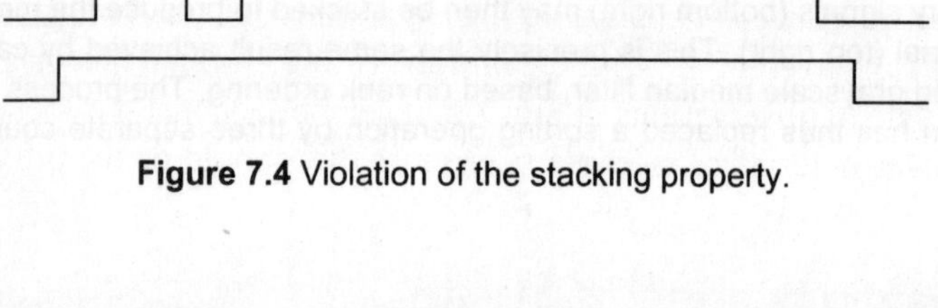

Figure 7.4 Violation of the stacking property.

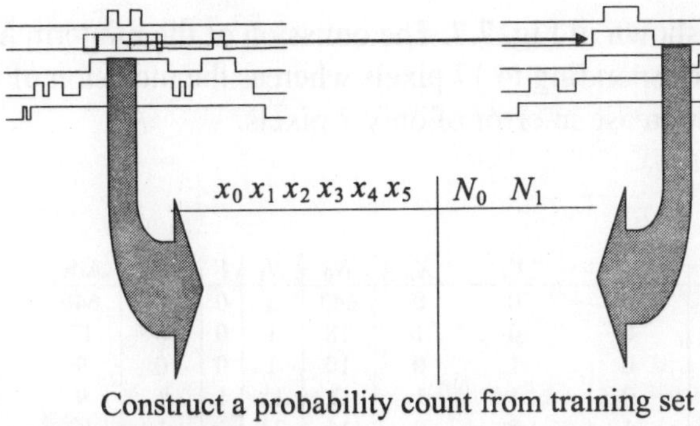

Figure 7.5 Stack filter design: counts are aggregated over all threshold levels.

ber of counts (N_0 and N_1) of the corresponding pixel in the ideal image is recorded as either 0 or 1. The counts are aggregated over every threshold level. As in the binary case, the optimum filter is found by setting the output value to 0 or 1 according to whether N_0 or N_1 is larger for each input combination.

However, a little caution is required here as there is no guarantee that this approach will result in a filter based on a PBF. An example of stack filter design is shown in Fig. 7.6. The noisy and ideal signals comprising the training set have been thresholded and a 3-point sliding window is used to compile the table of observations. The filter output is determined for each input combination depending on whether N_0 or N_1 is larger.

However, the function arising from this approach cannot be used as it will violate the stacking property which may in turn result in an output signal stack containing "holes". Instead of using the non-PBF arising from this process it is necessary to convert it to a PBF. Ideally the closest PBF, i.e., the one that causes the smallest increase in error, must be used. Tabus et al. devised a technique to convert a non-PBF to the closest PBF, resulting in the smallest increase in error.[3,4]

The technique is based on the understanding that there are two ways to remove complemented terms from binary functions:

- The minterms that cause the complemented terms in the non-PBF may be removed from the expression, or
- Further (mirror) minterms may be added to combine with the complemented terms and remove the necessity for negation.

In the example shown in Fig. 7.6, the problem is caused by the minterm $\overline{X}_0\overline{X}_1 X_2$. The solution is either to remove it from the function or include an additional minterm $X_0\overline{X}_1 X_2$ to combine with the one above and eliminate the negation. Referring to the table of observations in Fig. 7.6, it can be seen that the cost of the second option is lower in terms of increase in error so this should be the preferred option.

This comparison is shown in Fig. 7.7. The omission of the minterm $\overline{X}_0\overline{X}_1X_2$ would result in an error corresponding to 17 pixels whereas the inclusion of the matching minterm causes an increase in error of only 7 pixels.

Num	X_2	X_1	X_0	N_0	N_1	F	PBF	Adv
0	0	0	0	643	3	0	0	640
1	0	0	1	18	1	0	0	17
2	0	1	0	10	1	0	0	9
3	0	1	1	2	11	1	1	9
4	1	0	0	16	33	1	1	17
5	1	0	1	31	24	0	1	7
6	1	1	0	10	12	1	1	2
7	1	1	1	2	87	1	1	85

(a)

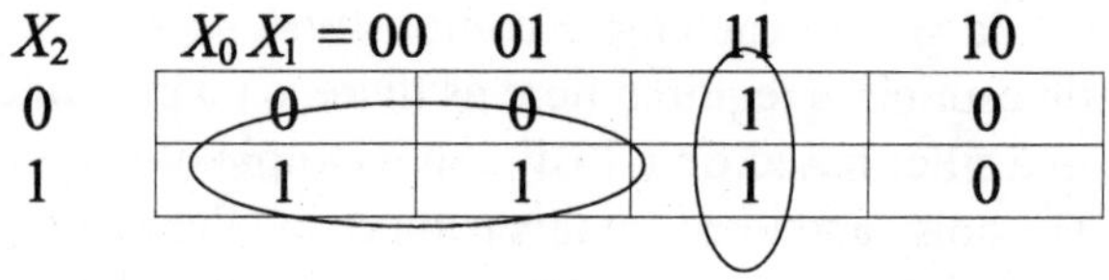

(b)

Figure 7.6 Ensuring a positive Boolean function. From the above observations (a), the optimum F is $F = X_0X_1 + X_1X_2 + \overline{X}_0X_2$. From the K MAP (b), it can be seen that F reduces to $F = X_0X_1 + \overline{X}_0X_2$, which is not a PBF. The problem is caused by term $\overline{X}_0\overline{X}_1X_2$.

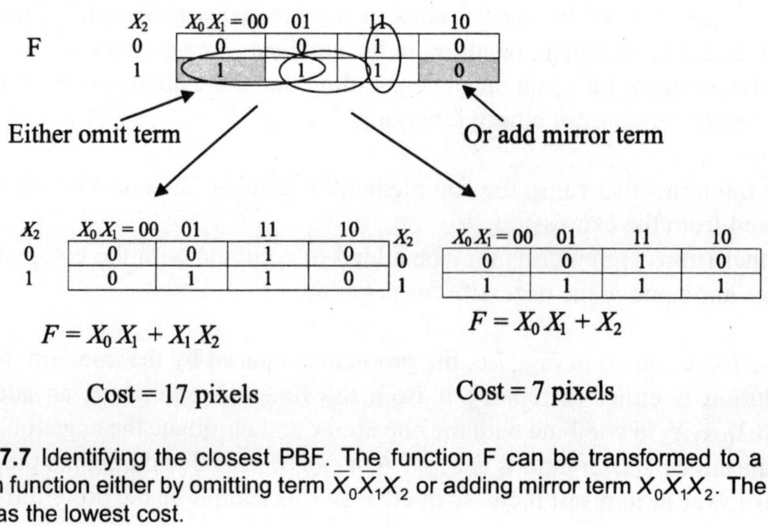

Figure 7.7 Identifying the closest PBF. The function F can be transformed to a positive Boolean function either by omitting term $\overline{X}_0\overline{X}_1X_2$ or adding mirror term $X_0\overline{X}_1X_2$. The latter of these has the lowest cost.

It is possible for a stack filter to be designed to use a different binary filter ψ_t for every threshold level. Care must be taken that the stacking property is not violated, so a constraint must be placed on the filters so that the ordering

$$\psi_{m-1} \leq \ldots\ldots \psi_{t+1} \leq \psi_t \leq \psi_1 \tag{7.7}$$

is preserved for every possible input. Dougherty calls this property *consistency*.[5]

The use of different filters at each threshold level can theoretically result in improved results since each filter can more closely model the required behavior at each level. However, if the total amount of training data is insufficient to be divided among the design of many filters, it can lead to worse results overall because of increased estimation error. Using the same filter at every level may be a compromise in terms of the different effects required at each level. However, the larger training set resulting from aggregating the data over all threshold levels can lead to a filter with a lower estimation error and better overall performance.

Stack filters can give excellent results for certain types of problems. Figure 7.8(a) shows a training set containing a noisy astronomical image and an ideal version. A stack filter was trained on these images and then applied to the image shown in Fig. 7.8(b). The noise is very severe in this type of data, and the stack filter does a good job of removing it. For comparison, Fig. 7.8(c) shows the two noisy images after filtering with Paintshop Pro, version 7 using the despeckle option. It can be seen that it makes little impression on the speckle. This is hardly surprising because it is operating without the benefit of an ideal image and is hence producing a general-purpose despeckle filter.

As can be seen above, given the right problem, stack filters can produce excellent results. They do, however, have strict limitations. Their processing structure treats the signal at each threshold level as an independent entity, and there is no communication path between levels. A training set with a brightness or contrast difference between the noisy and ideal image would confuse the filter and lead to poor results. Stack filters cannot detect objects or shapes because they are increasing filters, neither can they shift brightness levels. For more difficult problems it is necessary to link the threshold levels and use more complex filters. This is covered in the following sections.

7.2 Grayscale Morphology

All types of stack filters may be implemented through mathematical morphology and result in grayscale structuring elements that have vertical sides and flat tops. The process therefore acts on each threshold level independently. A more general type of filtering allows the structuring elements to take a shape of varying cross-scctions such as a triangle or cone. This has the effect of linking the processing across the threshold levels.

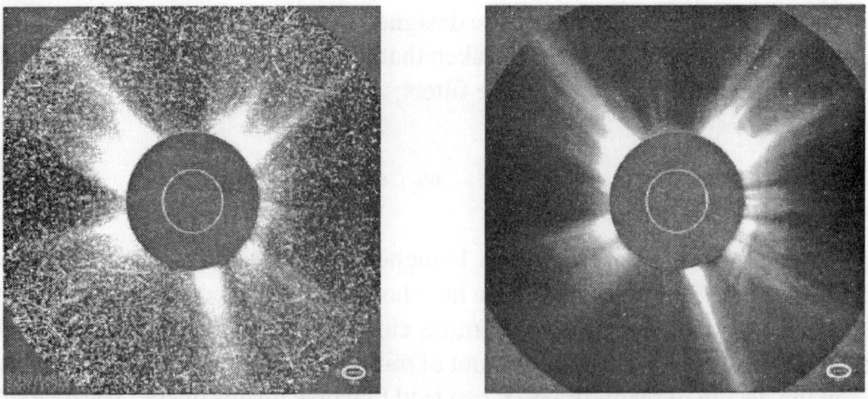

Figure 7.8(a) Training set of a noisy and ideal image. These images were used to design the stack filter applied below.

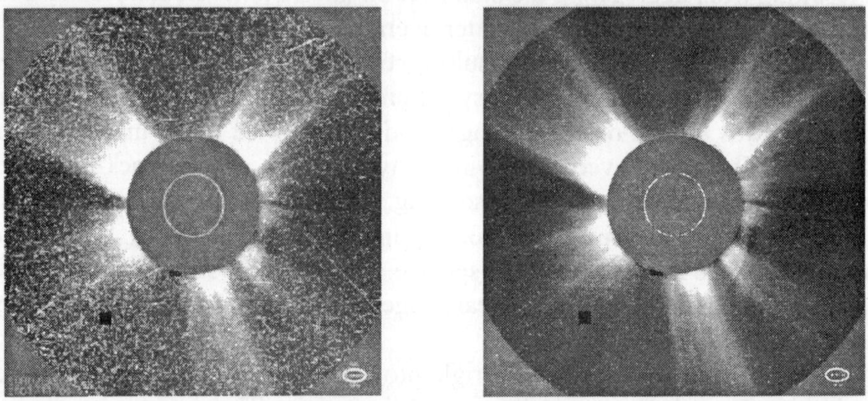

Figure 7.8(b) Noisy image and the filtered image resulting from application of the stack filter designed on training set above.

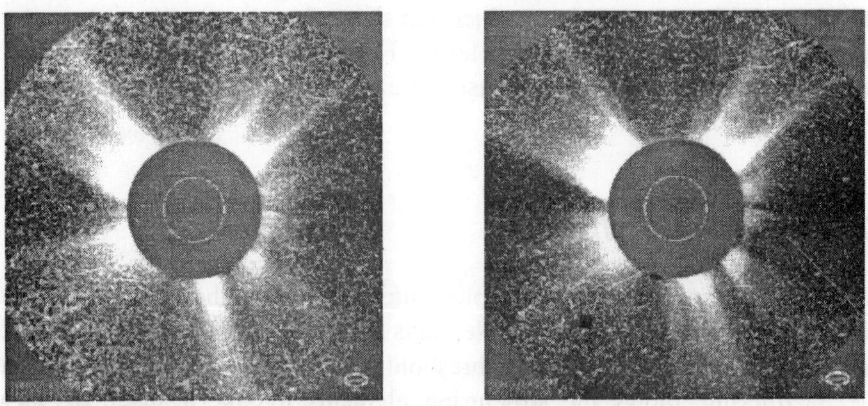

Figure 7.8(c) The two noisy images filtered using Paintshop Pro despeckle program. These standard filters can make little improvement with this type of noise.

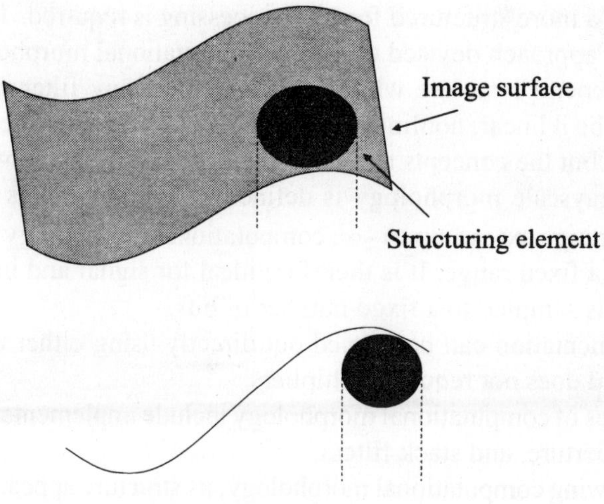

Figure 7.9 Grayscale morphology.

A commonly used shape of structuring element is a sphere or ball. Dilation and erosion operations may be effected by allowing the ball to roll over or under the surface and plotting the locus of the center of the ball. Similarly, openings and closings may be found by recording the places swept out by the surface of the ball as it rolls under or over the surface, respectively. Figure 7.9 shows an example of a spherical structuring element beneath the surface of a grayscale image. Such filters are effective at removing isolated noise spikes in grayscale images.

Grayscale morphology certainly represents a more general type of processing than stack filters. However, erosions and dilations by single structuring elements rarely achieve acceptable results. The simple design method for binary and stack filters does not readily generalize to grayscale filters. Design of grayscale structuring elements has so far only been achieved through iterative processes such as genetic algorithms.[6]

Grayscale morphology uses the same structuring elements at every level. There is therefore one further stage of generalization which allows the implementation of any type of filter, linear or nonlinear. This has been formalized by Dougherty as the concept of computational morphology.[7,8,9] It is an overall framework based on thresholding, and a brief introduction is given below.

7.3 Computational Morphology for Beginners

Since all digital imaging processes may be programmed or implemented in hardware, it is therefore possible to filter grayscale images by forming logical functions of all of the bits of the input. However, for a 3 × 3 filter with 8-bit data, this would result in a function of 72 bits. (Actually it would be 8 functions of 72 bits, one function for each of the output bits.)

In practice, a more structured form of processing is required. The most powerful and flexible approach devised to date is computational morphology.

This is a general structure which can implement any filter defined within a given window, be it linear, nonlinear, increasing, etc. The examples presented here use 1D signals, but the concepts extend to images in a straightforward way.

Whereas grayscale morphology is defined in the continuous domain and requires a signal range extending to $-\infty$, computational morphology works with discrete data over a fixed range. It is therefore ideal for signal and image processing where the data is sampled to a fixed number of bits.

The implementation can be carried out directly using either discrete logic or comparators and does not require multipliers.

Special cases of computational morphology include implementation of grayscale morphology, aperture, and stack filters.

On first viewing computational morphology, its structure appears to be very similar to that of stack filters and has three main components: thresholding, elemental erosion, and stacking. Stacking and thresholding have already been described as part of the stack filter description but elemental erosion is a new concept unique to computational morphology.

7.4 Elemental Erosion

An elemental erosion $\boldsymbol{e}$ is a grayscale-to-binary operation with two grayscale inputs (one waveform I and one structuring element B_i) resulting in a single binary output T_i, i.e.,

$$T_i = I\,\boldsymbol{e}\,B_i.$$

It is similar to a standard morphological grayscale erosion in that it probes whether the structuring element B_i "fits" beneath the waveform I and returns a 1 at the locations where it fits and a 0 where it does not. Hence, a binary signal T_i is produced. In the same way as standard morphology, the structuring element B_i has a single reference point which indicates the precise location where the output is affected.

The main difference between the elemental and standard grayscale erosion is that the structuring element is only allowed to move horizontally. It cannot move vertically and is "anchored" to the x axis. It thus produces a binary rather than grayscale output. Figure 7.10 illustrates elemental erosion.

In practice, elemental erosion is carried out over a set of structuring elements B_i known as a kernel. An increasing grayscale-to-binary filter, based on elemental erosion by a kernel of structuring elements B_i can be formed as a maximum of elemental erosions (sum of products),

$$T = I\,\boldsymbol{e}\,B_1 + I\,\boldsymbol{e}\,B_2 + \ldots\ldots\ldots\ldots\ldots I\,\boldsymbol{e}\,B_N\,, \tag{7.8}$$

where + represents the logical OR operator.

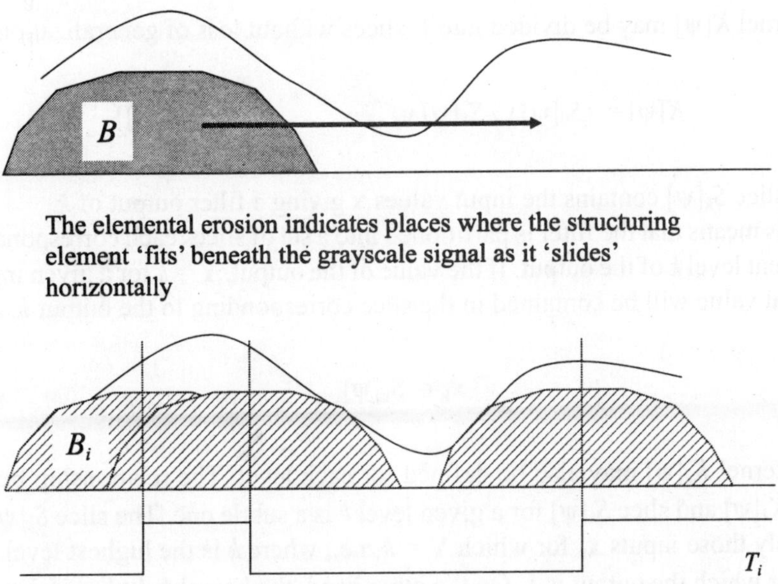

Figure 7.10 Brief illustration of elemental erosion. Unlike standard erosion, the structuring element is anchored to the *x* axis and slides horizontally. The output is binary and requires the entire structuring element to lie beneath the signal.

Note that Eq. 7.8 only applies to increasing filters. This is because it is based on a maximum (union) operation. Provided that the input exceeds a given level, it will cause the output to be 1. It is assumed that all inputs greater than this will also cause the output to be 1. However, for a nonincreasing filter, this will not be the case and the input must be further tested to determine if it falls within an *interval*. This concept is a generalization of the hit-or-miss transform discussed in Chapter 5. Further details are given in Dougherty.[5]

Although the output from the elemental erosion is binary, it may be used to model the behavior of a grayscale filter by representing a single output level *k*. A number of elemental erosions, each with a different kernel, are carried out in parallel—one for each level of the grayscale filter.

Consider the most general grayscale filter ψ. It can be represented by its kernel $K[\psi]$. This is very similar to a look-up table that returns an output value for any input combination.

If input **x** to the filter is a vector of *n* values all between 0 and $m-1$, and the output **Y** is a single grayscale value lying between 0 and $L-1$, this may be written as

$$\mathbf{x} \in \{0, 1, \ldots m-1\}^n \quad \text{and} \quad \mathbf{Y} \in \{0, 1, \ldots L-1\}. \tag{7.9}$$

The kernel $K[\psi]$ may be divided into L slices without loss of generalization.

$$K[\psi] = \{S_0[\psi] \cup S_1[\psi] \cup \ldots\ldots\ldots\ldots\ldots\ldots S_{L-1}[\psi]\}, \tag{7.10}$$

where slice $S_k[\psi]$ contains the input values $\mathbf{x}$ giving a filter output of k.

This means that the filter is partitioned into a set of slices each corresponding to a different level k of the output. If the value of the output, $\mathbf{Y} = k$ for a given input $\mathbf{x_k}$, then that value will be contained in the slice corresponding to the output k, i.e.,

$$\mathbf{x_k} \in S_k[\psi].$$

A kernel $K_k[\psi]$ may also be defined for each level. The relationship between kernel $K_k[\psi]$ and slice $S_k[\psi]$ for a given level k is a subtle one. The slice $S_k[\psi]$ contains only those inputs $\mathbf{x_k}$ for which $\mathbf{Y} = k$, i.e., where k is the highest level of the stack for which the output is 1. On the other hand, the kernel $K_k[\psi]$ contains all inputs $\mathbf{x_k}$ for which that level of the stack is 1, i.e., $\mathbf{Y} \geq k$.

In other words, the slice $S_k[\psi]$ contains inputs for which the output exactly corresponds to level k and the kernel contains inputs for which the output is level k or greater. The kernel $K_k[\psi]$ may therefore be written as

$$K_k[\psi] = \{S_k[\psi] \cup S_{k+1}[\psi] \cup \ldots\ldots\ldots\ldots\ldots\ldots S_{L-1}[\psi]\}. \tag{7.11}$$

Dougherty showed that any operation linear or nonlinear may be placed in the context of computational morphology.[7,8] The framework is shown in Fig. 7.11. The input signal is subjected to an elemental erosion by a set of kernels of structuring elements. The output from each elemental erosion produces a binary signal for the appropriate level of the output. All of these binary signals are stacked to produce the grayscale output signal.

Although a 1D signal is shown here, the principle may be extended to images in which the structuring elements correspond to windows of gray level values. Unlike stack filters, there is no thresholding of the input signal; the full grayscale signal is subjected to the elemental erosion for every level of the output. In the most general case, the kernel for every output level is different although there must be an ordering such that

$$K_1[\psi] \supseteq K_2[\psi] \supseteq \ldots\ldots\ldots\ldots\ldots\ldots K_L[\psi]. \tag{7.12}$$

This is a more general condition than the equivalent stacking property in stack filters required to preserve the stacking property of the outputs in computational morphology.

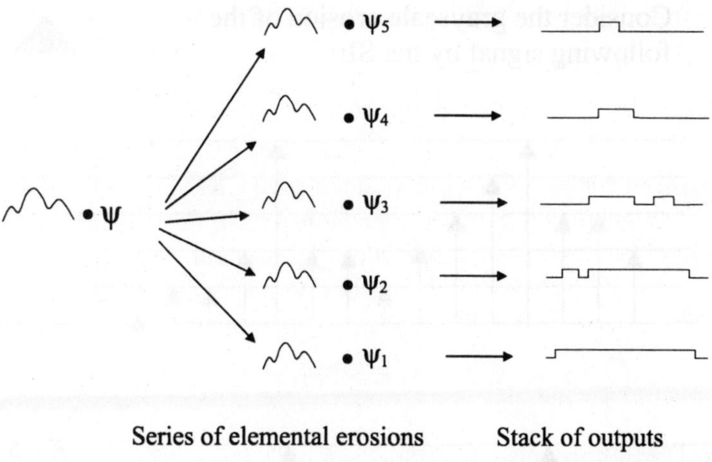

Figure 7.11 Computational morphology. The input grayscale signal is subjected to elemental erosion by the kernel for each level. This results in a grayscale signal represented as a binary output stack.

The most difficult part of computational morphology is determining the contents of each kernel of structuring elements. There can be a very large number of structuring elements in each kernel, and for an unconstrained filter the training data required to determine these can be impossibly large. A simple constraint that produces good practical filters is the aperture constraint. This results in the aperture filter, which will be described shortly.

Both grayscale morphological and stack filters are special cases of computational morphology. In grayscale morphology, the structuring elements in each kernel are the same shape and related by an offset:

$$\mathbf{x} + k \in K_k[\psi], \tag{7.13}$$

where the scalar value k is added to every component of $\mathbf{x}$.

This is better illustrated through an example. Consider the grayscale morphological erosion of the 5-level signal in Fig. 7.12 by a triangular structuring element. In the morphological erosion, the structuring element is placed as high as it will go while remaining under the signal.

To implement the above grayscale morphological filter through computational morphology, it is required to determine the five kernels $K_k[\psi]$ for $k = 1$ to 5.

The kernel $K_k[\psi]$ specifies the structuring element(s) for the elemental erosion which will generate the output at level, k.

These structuring elements are related by a simple scalar offset. That is, the SE to be applied at level 1 is (0,1,0). For level 2 it is shifted up by a value of 1, etc. The set of SE are shown in Fig. 7.13.

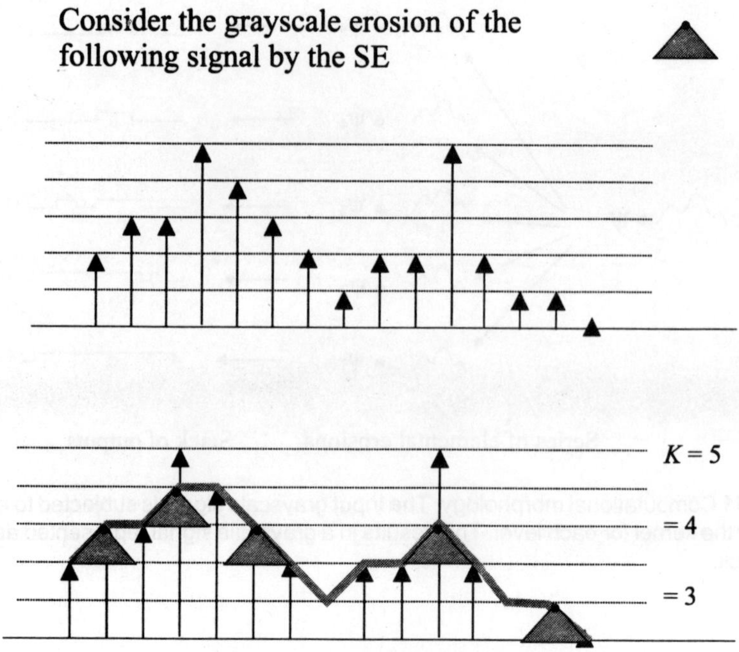

Figure 7.12 Greyscale erosion.

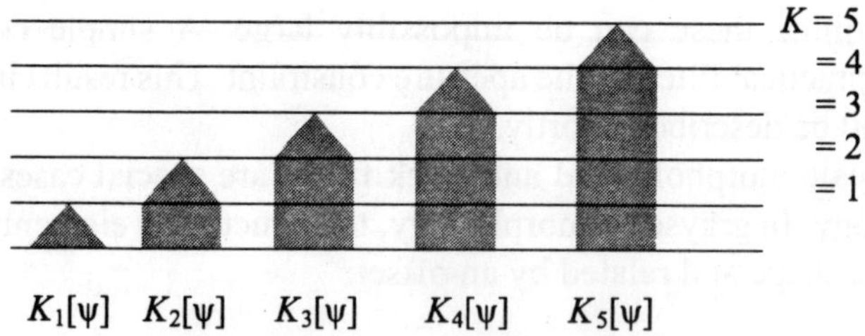

Figure 7.13 Filter kernels for grayscale erosion.

$$K_1[\psi] \text{contains } \mathbf{x} = (0, 1, 0)$$

$$K_2[\psi] \text{contains } \mathbf{x} = (1, 2, 1)$$

$$K_3[\psi] \text{contains } \mathbf{x} = (2, 3, 2)$$

……………………………

$$K_k[\psi] \text{contains } \mathbf{x} = (k - 1, k, k - 1) \tag{7.14}$$

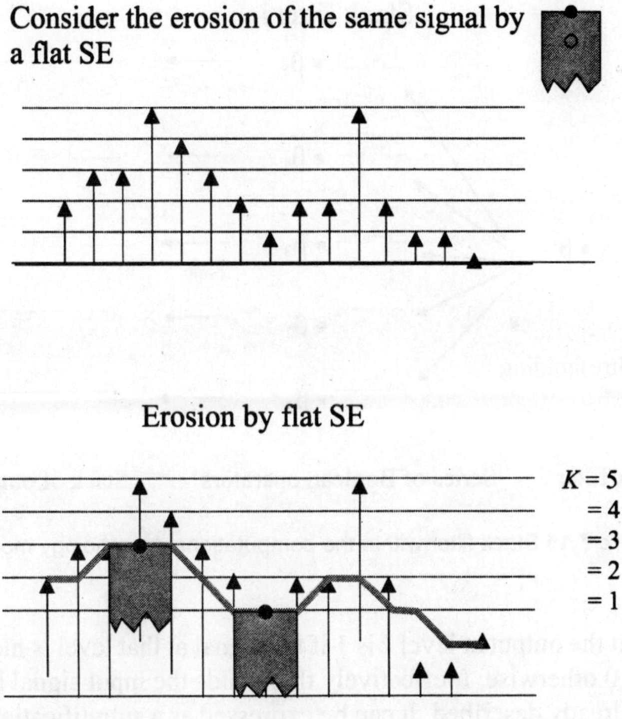

Figure 7.14 Stack filter example showing erosion by a "flat" structuring element.

So the grayscale morphological filter represents a constraint on the general filter as the contents of the kernels at each level are forced to be the same shape related through a simple offset.

Stack filters may also be put in the context of computational morphology and in this case the structuring elements are not only related by an offset but are constrained to be "flat" with vertical sides. Figure 7.14 shows an example of an erosion of a grayscale signal by a flat structuring element.

Notice that the flat structuring element is drawn with the bottom edge jagged. The only points that matter are the top surface so the SE could just as easily have been drawn as a horizontal line three points wide.

The kernel for each output level $K_k[\psi]$ consists of

$$k = 1 \quad \mathbf{x} = (1, 1, 1)$$

$$k = 2 \quad \mathbf{x} = (2, 2, 2)$$

$$\text{i.e.,} \quad \mathbf{x} = (k, k, k) \tag{7.15}$$

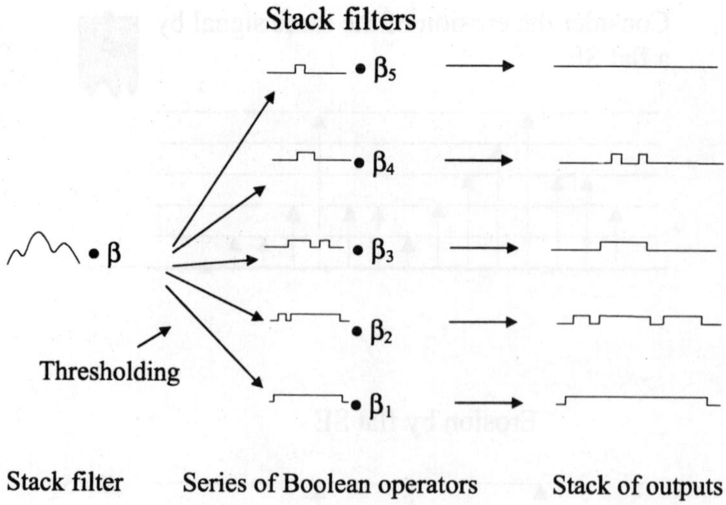

Figure 7.15 Stack filter within the computational morphology model.

This means that the output at level k is 1 if the signal at that level is more than three points wide or 0 otherwise. It effectively thresholds the input signal leading to the stack filter as already described. It can be expressed as a simplification of the computational morphology model with thresholding of the input signal and binary filtering of the different levels (Fig. 7.15).

7.5 Aperture Filters

As mentioned earlier, computational morphology is such a general framework that it can be difficult to design, i.e., it can be difficult to determine the filter kernel. This is because the number of possible input combinations is huge. For an 8-bit 1D signal, a 5-point window would have $2^{5\times8} = 2^{40} \approx 10^{12}$ inputs and so constraints are needed. One recently introduced constraint is the aperture constraint. In the same way that the window constraint limits the inputs to those falling within a finite spatial interval, the aperture constraint limits the inputs to a finite interval in amplitude. The signal is viewed through a rectangular window known as an aperture. The principle will be described in terms of 1D signals, but the concept extends readily to images where the aperture becomes a rectangular "box". The aperture slides along the signal and moves up and down to track the signal level. An example of a signal and aperture is shown in Fig. 7.16(a).

Aperture placement is an interesting topic and will be discussed in more detail later. Where a point of the signal lies beyond the top (or bottom) of the aperture, it is clipped to the highest (or lowest) value within the aperture window. Figure 7.16(b) shows the quantized samples in an aperture with five spatial points and seven quantization levels. Without further constraint, the aperture has $7^5 = 16807$ input

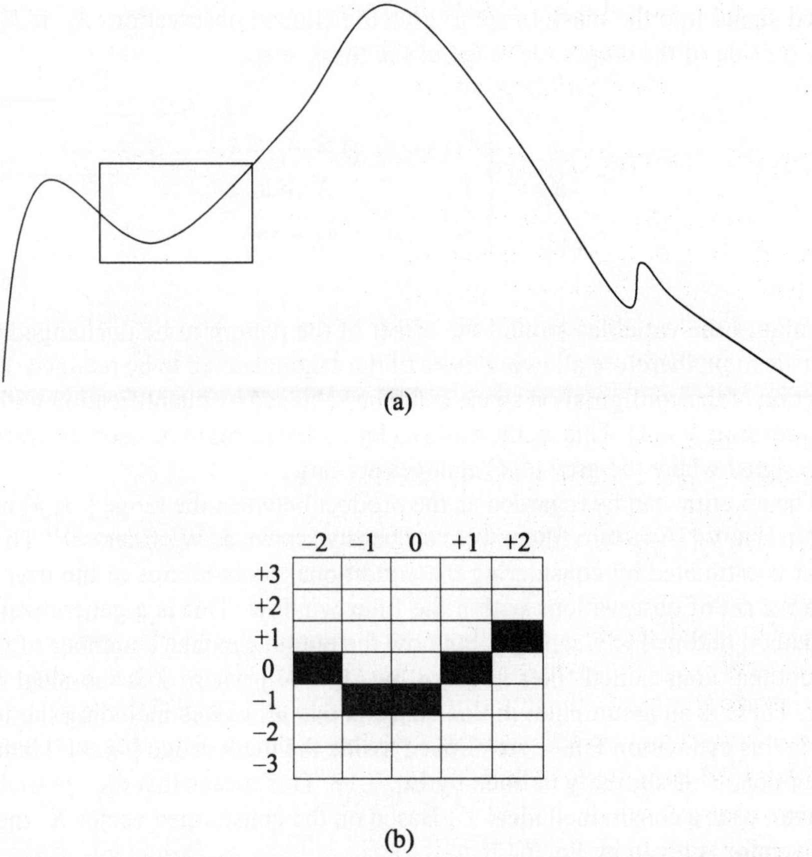

Figure 7.16 Aperture filter. (a) Aperture placed on the signal. (b) Quantized samples within the aperture.

combinations. As well as extending the windowing property to the signal amplitude, the aperture filter also extends the concept of translation invariance to the vertical direction since training data is combined from apertures placed at different levels.

In the same way, many of the input patterns within the aperture may be combined in training if they differ by only an offset. In this case, the aperture output value falls between + 3 and – 3. In theory, the output does not have to be clipped. Although the design procedure is developed with a clipped output Y^*, in the next few paragraphs it could theoretically work with the unclipped values of Y.

Like all filters in this class, aperture filters are designed using a training set consisting of ideal and corrupted versions of the signal. A window constrained in both the amplitude and the domain of the signal is used to collect the data. The constraint in the range means that certain signal values that fall outside the amplitude range $K = [-k, k]$ of the window are "clipped" to the top or the bottom value of the window range. Equation 7.16 gives the function mapping each point X_j of the original observed signal into the mask range to give the clipped observations X_j^* if X_j is initially outside of the amplitude range of the mask shape.

$$X_j^* = \begin{cases} X_j, & -k \le X_j \le k, \\ k, & X_j > k, \\ -k, & X_j < -k. \end{cases} \tag{7.16}$$

This allows the variables around the offset of the pattern to be unchanged by the quantification, therefore allowing more of the original detail to be retained. Further reduction of the configuration space could be achieved by quantification within the aperture range $[-k, k]$. This method allows larger apertures to be used to cover areas in the signal where the gray level changes are large.

The aperture can be regarded as the product between the range $[-k, k]$ and the domain $[-w, w]$. Aperture filters were originally known as W_K filters.[10,11] The filter output is estimated by considering the conditional probabilities of the true signal given the set of observations within the filter window. This is a generalization of the method outlined in Chapter 2, but now the output can take a number of values. The optimal constrained filter is given by $E[Y \mid \mathbf{X}^*]$ where Y is the ideal output value. There is an assumption in this analysis that all values including the ideal Y used in this estimation fall or are clipped inside the mask range $[-k, k]$. Under this assumption, Y^* is similarly defined by Eq. 7.17. This means that the optimal MSE estimator uses a constrained ideal Y^*. Based on the constrained vector $\mathbf{X}^*$ the optimal operator is given by Eq. 7.17:

$$\psi_A = E[Y^* \mid \mathbf{X}^*]. \tag{7.17}$$

As with any constraint, there is an associated cost. In this case, there is an error (resulting from range constraint) for using the aperture filter ψ_A instead of a window filter that is not constrained in amplitude ψ_W. This error, in terms of mean-absolute error, is given by Eq. 7.18. Further details are given in Hirata.[12]

$$\Delta(\psi_A, \psi_W) = E[|Y - E[Y^* \mid \mathbf{X}^*]|] - E[|Y - E[Y \mid \mathbf{X}]|]. \tag{7.18}$$

In order to estimate the conditional probabilities, the aperture has to be positioned in the signal space. Placement of the window can be done in various ways, but the most important consideration is the reduction of the number of points falling outside the window range. Examples of placement are explored in Hirata et al.[13] These methods involve referencing the aperture to the observed value or the median of the observed pattern in the domain window. In general, the best aperture placement strategy is the one that gives the closest estimate of the output. This will vary depending on the nature of the problem. In removal of impulsive noise, it could be the median. The aperture would then act as a "correction" to the median. The resulting filter should therefore never be worse than the best known to date, but may well be better.

A number of variations on the aperture filter have been introduced including multimask and two types of multiresolution approaches. Hirata et al. used a window that was defined at several resolutions.[13] At the finest resolution, every location in the window was used. At coarser resolutions, these locations were combined to give a window that covers a larger area with a smaller number of cells. This is illustrated in Fig. 7.17(a). The filter switches between the scales as follows: When a new pattern is encountered, the training set is checked to determine if that pattern was observed a given number of times at the finest resolution. If it was, the output is determined from this data. If it was not observed at the finest resolution, the next resolution is checked. It is more likely that a pattern will have been observed at a coarser resolution because the search space is much smaller. This process proceeds, and an output is formed using the finest resolution for which sufficient training data is available.

A different type of multiresolution approach was introduced by Green et al.[14] This work uses a single H-shaped window as shown in Fig. 7.17(b). The resolution of the window becomes coarser towards the edges. The principle here is that the fine details are captured at the center of the window and that the overall signal shape is captured by the coarser cells at the extremes of the window. The latter al-

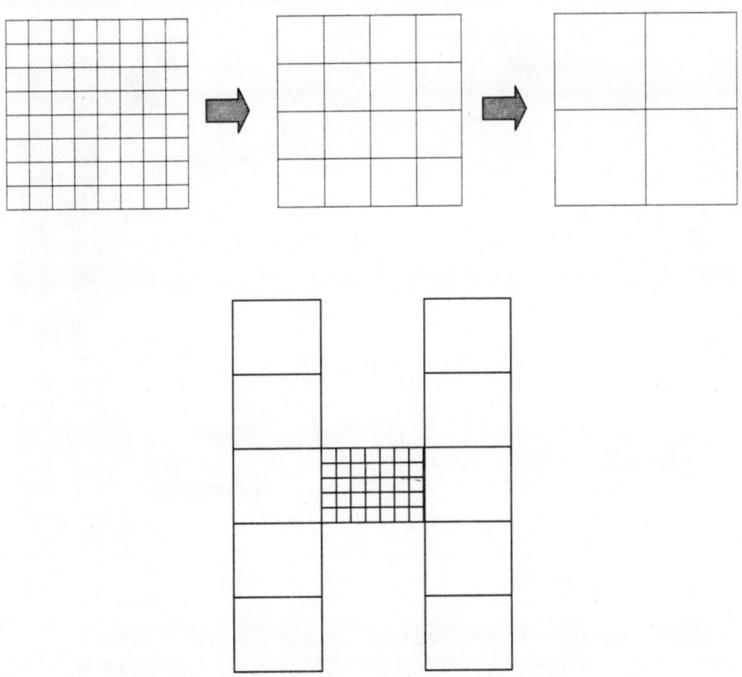

Figure 7.17 Multiresolution aperture filter.

lows the overall signal shape to be taken into account without increasing the search space drastically. For realistic-sized training sets, results show that the H-shaped window gives better results than either the smaller area in the center or the overall H window at full resolution.

An example is given for 1D apertures comparing two full-resolution apertures and an H-shaped multiresolution aperture. Each operator was designed to denoise a signal in which 10% of the points have been corrupted with Gaussian noise of variance 5. The test was carried out using 60 training signals, each of length 1024. The first aperture is an H-shaped aperture ψ_H where all of the cells are retained at the finest resolution. The second aperture is a "standard" aperture ψ_S. This occupies just the central portion of the H-shaped aperture. The third aperture is the multiresolution aperture ψ_M. It is produced by taking the H-shaped mask and mapping the groups of cells furthest from the center into single large cells. It was arranged that the total number of cells was the same for the second and third aperture.

Figure 7.18 shows the three apertures and gives the plots of MAE after filtering compared to the number of training examples. It can be seen that for the amount of training data used, the standard aperture performs much better than the H-shaped aperture. This is because there are far fewer patterns to be optimized in the standard aperture and so it has a much lower estimation error, i.e., $\varepsilon[\psi_H, n] > \varepsilon[\psi_S, n]$.

Figure 7.18 also gives the MAE plot for the multiresolution aperture. The multiresolution aperture combines estimates from a number of high-resolution patterns, it therefore gives a better estimate of the ideal signal. It also spans a larger area without increasing the size of the search space. It further gains an advantage

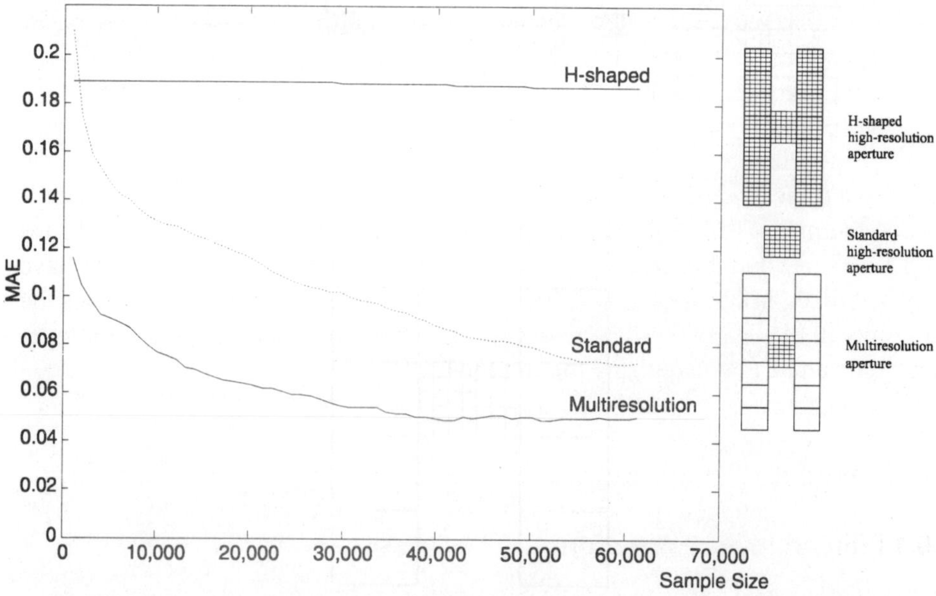

Figure 7.18 Comparison of H-shaped, standard, and multiresolution apertures.

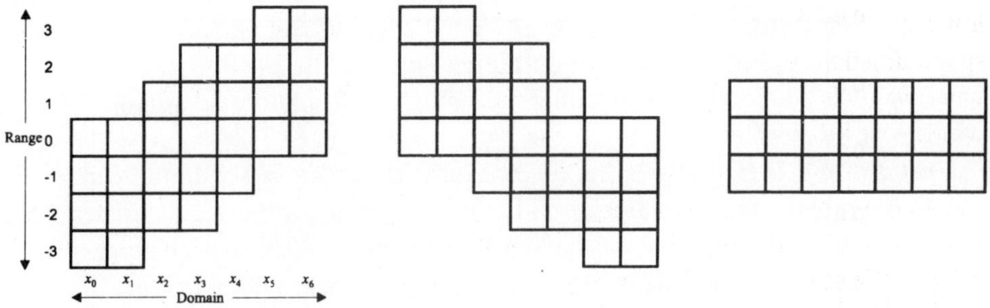

Figure 7.19 Multimask aperture filters.

from grouping similar patterns together, therefore unseen patterns will have a better estimate than if the signal was trained by the standard aperture.

Another approach is the multimask approach[15] where the overall filter window is decomposed into a number of differently shaped subwindows known as masks. The masks are designed to represent commonly occurring shapes within the signal. An example of the masks are given in Fig. 7.19.

The multimask filter was compared to the standard aperture and the median filter in terms of its ability to remove random noise. The training and test set were as described above but the number of examples was increased from 1024 to 61,440 training pairs. This test was carried out because the median can be used to place the aperture to collect the observations. The results show how, by selecting the output by conditional expectation, the multimask filter improves the decision made by the median filter. Although the two filters are effective for noise removal, the median filter tends to remove small features whereas the multimask learns to preserve by estimation from the training set. The domain of the multimask and standard aperture filters was 7 × 7 points and the median filter was also computed over seven points.

Figure 7.20 shows the error performance of the multimask aperture filter (in this case it is the MSE that is used) compared with both the single mask aperture filter and the median filter. It can be seen that the multimask aperture outperforms both filters. Although neither the multimask nor the single mask apertures have fully stabilized in terms of estimation error, even at 60,000 training examples the difference in MSE between these two designs is large enough to show the improvement that can be obtained using the multimask design.

7.6 Grayscale Applications

7.6.1 Film archive restoration

Morphological filters may be trained for removal of noise from film and video footage. This has been successfully applied to old film archive restoration using

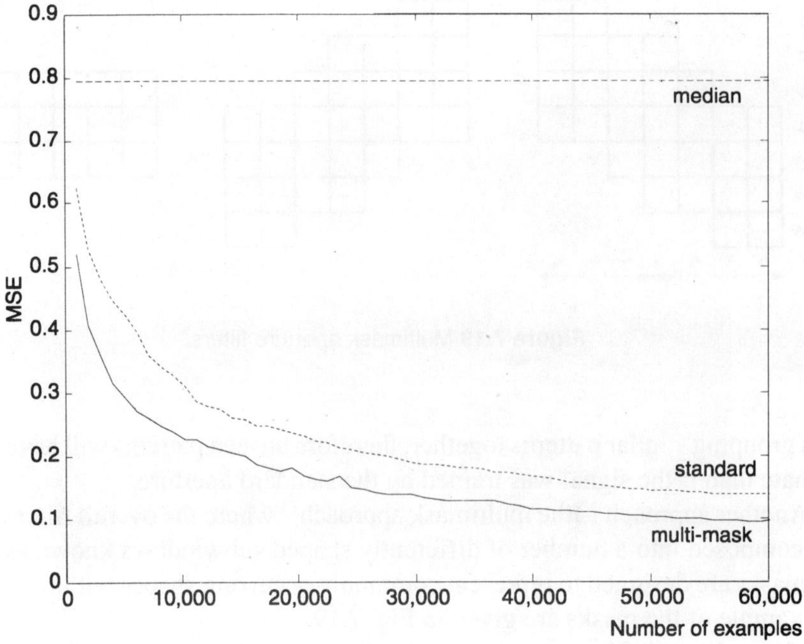

Figure 7.20 MSE plot comparing the multimask aperture, the standard aperture, and the median filter.

spatio-temporal filters. The filter kernel contains structuring elements which are 4D. That is, they exist in two intraframe dimensions of space (vertical and horizontal), one interframe dimension of time, plus they have intensity values. In this case, genetic algorithms were used to optimize the parameters of the filter over a training set. The training set was created by selecting relatively clean parts of the footage and pasting in noise blotches. Comparisons were made with non-training set techniques such as the optimization of image quality parameters and the training set. Procedures were always found to be superior provided that the training set was representative of the noise and image. Further details can be found in Hamid,[16] Kraft,[17] and Marshall.[18]

Figure 7.21 shows an example of old film restoration where noise "blotches" have been removed without damaging the fine image structure.

7.6.2 Removal of sensor noise

An unsightly artifact of low-light imaging is the appearance of sensor noise. This produces speckle on the image. Figure 7.22 shows an example of a frame taken from low-light footage. The lower image has been despeckled using a morphological filter. In comparison, commercial packages such as Paintshop Pro had little success in restoring this frame.

Figure 7.21 Old film restoration. The top image has had black noise "blotches" removed using a spatio-temporal morphological filter. Notice that the fine structures remain intact.

(a) Image with sensor noise

(b) filtered image

Figure 7.22 An example of removal of sensor noise caused by low lighting conditions. (a) shows the image with sensor noise; (b) displays the filtered image.

It was found that for video sequences as opposed to still images it was necessary to employ spatio-temporal filters to avoid motion artifacts.

7.6.3 Image deblurring

A further problem in film footage is that of blurring resulting from motion and from autofocus cameras during panning when the distance to the subject has not stabilized. It is very easy to create training data for blur problems simply by low-pass filtering footage with sharp detail and using this as the noisy data with the original footage as the ideal data.

Figure 7.23 shows an example of image deblurring with an aperture filter. The aperture was trained on the deliberately blurred image of a lab. Figure 7.23(a) shows a blurred image and Fig. 7.23(b) shows the same image after application of the deblurring aperture. It can be seen that the clock is distinctly sharper.

(a)

(b)

Figure 7.23 Example of image deblurring implemented through aperture filtering. The top image is slightly blurred and the lower one has been sharpened by an aperture filter deblurring process.

7.7 Summary

This chapter has explained how the binary restoration techniques from the earlier chapters may be extended to apply to grayscale processing. The extension may take place using a number of techniques representing different stages of generalization. At each stage there is the usual trade-off between constraints and training.

Stack filters, for example, simply duplicate the binary process over a number of threshold levels. Given the right type of problem, they give excellent results. They are highly constrained and not applicable to many types of problems. Generalization to grayscale and computational morphology results in a much more flexible and powerful type of processing, but presents serious problems in terms of training. The simple statistical approach used in the binary case is no longer practical, and a combination of iterative methods and imaginative constraints is required.

An emerging partial constraint produces the aperture filter that has shown itself to be useful for a number of practical problems. Further variations on the aperture filter such as multimasking and the two different multiscale approaches have successfully increased the region of support of the filter without a corresponding explosion in the size of the search space—so much so that direct statistical design techniques, rather than iterative search methods, has become feasible.

Several grayscale examples have been presented including film restoration, deblurring, and despeckle to demonstrate the widespread range of practical imaging problems that may be addressed with these methods.

One aspect of the work not yet considered in detail is the practical implementation. This is the subject of the next chapter.

References

1 P. D. Wendt, E. J. Coyle, and N. C. Gallagher, "Stack filters," *IEEE Trans. Acoustics, Speech, and Signal Processing*, **34**(4), 898–911 (1986).

2 F. Y. Shih and O. R. Mitchell, "Threshold decomposition of gray scale morphology into binary morphology," *IEEE Transactions on Pattern Analysis and Machine Intelligence*, **11**(1), 31–42 (1989).

3 I. Tabus, D. Petrescu, and M. Gabbouj, "A training framework for stack and boolean filtering–fast optimal design procedures and robustness case study," *IEEE Transactions on Image Processing, Special Issue on Nonlinear Image Processing*, **5**(6), 809–826 (1996).

4 D. Petrescu, I. Tabus, and M. Gabbouj, "Optimal design of boolean and stack filters and their application in image processing," in *Nonlinear Model-Based Image/Video Processing and Analysis*, C. Kotropoulos and I. Pitas (eds.), Wiley, New York, 15–58 (2001).

5 E. R. Dougherty and J. Barrera, "Computational gray-scale image operators," in *Nonlinear Filters for Image Processing*, E. Dougherty and J. Astola (eds.), 61–98, SPIE Press, Bellingham, WA (1999).

6 N. R. Harvey and S. Marshall, "GA optimisation of multidimensional gray-scale soft morphological filters with applications in archive film restoration," in *Mathematical Morphology and its Applications to Signal Processing*, ISMM 2000, Palo Alto (2000).

7 E. R. Dougherty and D. Sinha, "Computational mathematical morphology," *Signal Processing*, **38**, 21–29 (1994).

8 E. R. Dougherty and D. Sinha, "Computational gray-scale mathematical morphology on lattices (a comparator-based image algebra)—Part I: Architecture," *Real-Time Imaging,* **1**(1), 69–85 (1995).

9 E. R. Dougherty and D. Sinha, "Computational grey-scale mathematical morphology on lattices (a comparator-based image algebra)—Part II: Image operators," *Real-Time Imaging*, **1**(4), 283–295 (1995).

10 J. Barrera and E. R. Dougherty, "Representation of grayscale windowed operators, mathematical morphology and its applications to image and signal processing," in *Computational Imaging and Vision*, vol. 12, H. J. Heijmans and J. B. Roerdink (eds.), Kluwer Academic Publishers, Dordrecht, 19–26 (1998).

11 R. Hirata, E. R. Dougherty, and J. Barrera, "Optimal range-domain window filters," *Proc. SPIE*, **3646**, 38–45 (1999).

12 R. Hirata, E. R. Dougherty, and J. Barrera, "Aperture filters," *Signal Processing*, **80**, 697–721 (2000).

13 R. Hirata, M. Brun, J. Barrera, and E. R. Dougherty, "Multiresolution design of aperture filters," *Mathematical Imaging and Vision*, **16**(3), 199–222 (2002).

14 A. C. Green, E. R. Dougherty, S. Marshall, and D. Greenhalgh, "Optimal filters with multiresolution apertures," *J. Math. Imaging Vis*., **20**(3), 237–250 (2004).

15 A. C. Green, E. R. Dougherty, S. Marshall, and D. Greenhalgh, "Design of multi-mask aperture filters," *Signal Processing*, **83**(9), 1961–1971 (2003).

16 M. S. Hamid, S. Marshall, and N. Harvey, "GA optimisation of multidimensional gray-scale soft morphological filters with applications in archive film restoration," *IEEE Trans. Circuits and Systems for Video Technology*, **13**(5), 406–416 (2003). See also **13**(7), 726 (2003).

17 P. Kraft, N. Harvey, and S. Marshall, "Parallel genetic algorithms in the optimization of morphological filters: a general design tool," *J. Electron. Imaging*, **6**(4), 504–516 (1997).

18 S. Marshall, N. Harvey, and D. Greenhalgh, "Design of morphological filters using genetic algorithms," *EUSIPCO 2000*, Tampere, Finland (2000).

Chapter 8
Grayscale Implementation

This chapter considers some of the implementation issues that are encountered in processing grayscale images. These issues fall into two main areas: grayscale training issues and grayscale hardware implementation.

8.1 Grayscale Training Issues

8.1.1 Envelope filtering

It was seen in Chapter 4 that training of filters to deal with real-world problems is a difficult task. A balance must be struck between the dimensionality of the training set and the size of the search space. The training task may be simplified by limiting the complexity of the problem through the application of a constraint. This leads to an increase in error due to the addition of a constraint error term. However, this may be more than offset by the reduction in estimation error. Estimation error resulting from inadequate training of filters can be very severe and can even result in filters that actually increase the error.

Equation 4.2 illustrated the trade-off involved when a filter is constrained by reducing its window size. A similar trade-off applies to other types of constraints. A recently introduced constraint that has been shown to be very valuable for the type of nonlinear filters described in this book is the use of envelopes. The concept of envelopes is not new in itself, and it has been used in other areas of signal and image processing.[1,2,3,4] However, an example of its application to grayscale filter design was introduced by Brun et al.[5]

While other constraints such as the window constraint have limited the input to the filter, and constraints on the class of functions have limited the processing of the input, envelope constraints directly constrain the output of the filters. The designed filter processes the data and produces an output that is then constrained to lie between a lower and upper bound, α and β, respectively. Figure 8.1 shows an example of a signal ψ constrained to an envelope resulting in the signal ψ_{con}. Where ψ

falls within the envelope, the original value is retained. However, where it falls above β or below α, its value is trimmed to the envelope's extremities. This is set out in the equation below.

$$\psi_{con} = \begin{cases} \alpha & \text{if} \quad \psi < \alpha, \\ \psi & \text{if} \quad \alpha \le \psi \le \beta, \text{ and} \\ \beta & \text{if} \quad \psi > \beta \end{cases} \tag{8.1}$$

The envelope represents the upper and lower bounds of the expected output from the filter with the optimum filter output ψ_{opt} ideally lying somewhere in between. As will be seen in later examples, the envelope is usually formed from a simple combination of filters. The effect of an envelope constraint may either reduce or increase the filter error. The search space is made smaller, so the filter output may be better estimated from a smaller training set. However, there is an increase in constraint error since the filter is limited in its range of outputs. Where it lies outside the envelope there will be an error introduced by restricting the output to the closest edge of the envelope, either β or α. The reader should remember that in general the quantity being trimmed is not the output from the optimum filter ψ_{opt}, but an estimate derived from a finite number of training samples $\psi_{opt,N}$ where N is the number of samples. A well-designed envelope will prevent very large errors from occurring.

Brun et al.[5] have produced mathematical proofs to show that the optimum filter with output lying between β and α may be obtained by determining the optimum filter without the envelope constraint and trimming it to the envelope. This simpli-

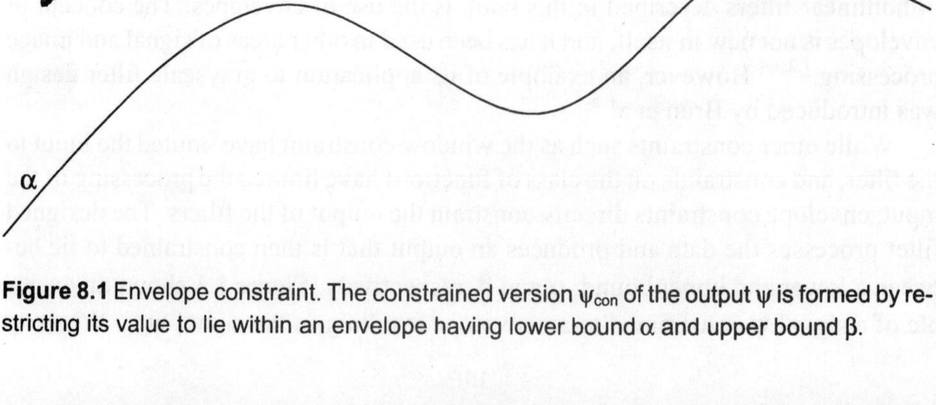

Figure 8.1 Envelope constraint. The constrained version ψ_{con} of the output ψ is formed by restricting its value to lie within an envelope having lower bound α and upper bound β.

fies the design strategy. They also show that when the optimum filter output lies within the envelope, the constraint can only be beneficial.

An example demonstrating the benefit of envelope filtering taken from the above paper is presented here. Figure 8.2(a) shows a corrupted image that has been created by adding both 10 percent salt-and-pepper noise and a series of horizontal

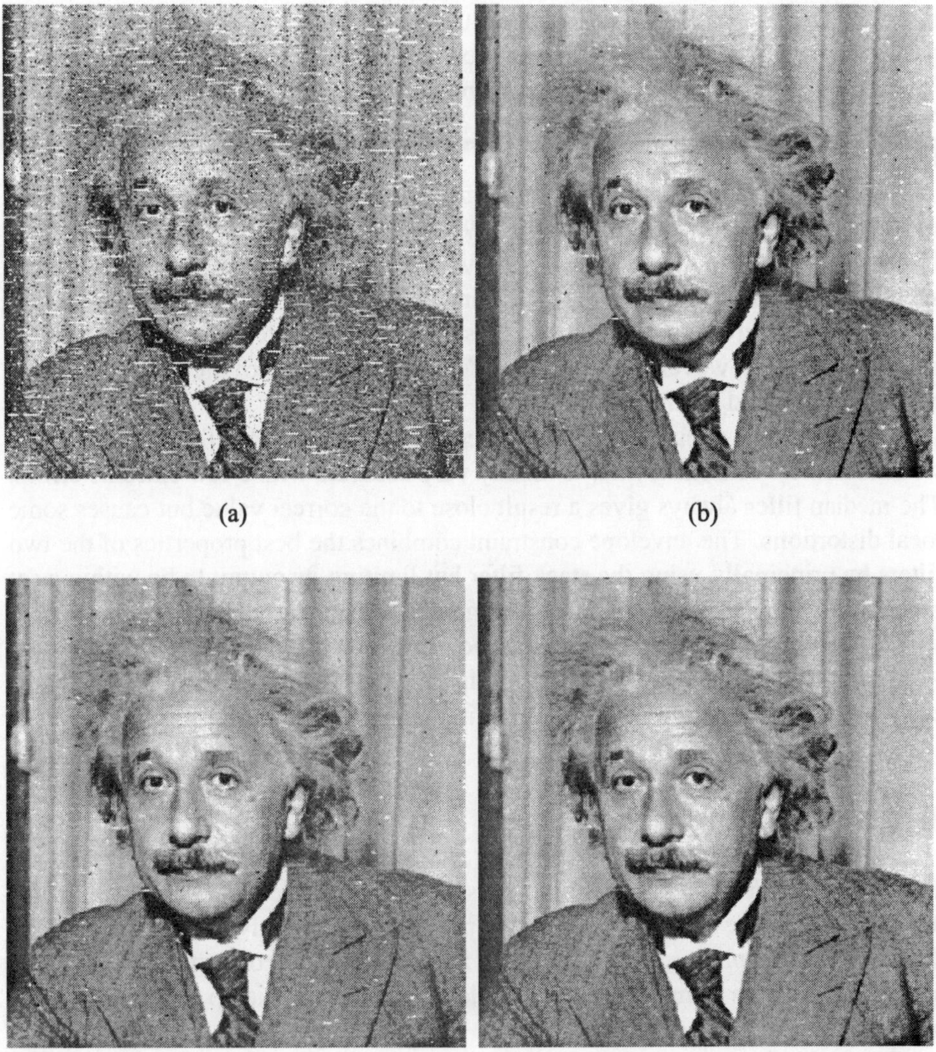

(a) (b)

(c) (d)

Figure 8.2 Envelope filtering example. The above figures show the benefit of envelope filtering. The noisy image is shown in (a) and its restoration with a 17-point stack filter is shown in (b). Envelope filtering using openings and closings by a cross results in the image in (c). Envelope filtering to within a fixed distance of the median is shown in (d). The MSE for each image is (a) 1912; (b) 106; (c) 79; and (d) 55, respectively.

line segments with parameters drawn from a normal distribution. The image was restored using a stack filter applied within a 17-point window both with and without envelope filtering. The result of applying the stack filter alone is shown in Fig. 8.2(b).

Two different envelopes were used. In the first example, the upper bound β, and lower bound α, of the envelope were set to an opening-closing and a closing-opening of the filtered image. The structuring element was a 3×3 cross dilated by itself. This means that any extreme values of the image remaining after stack filtering were trimmed off. The result is shown in Fig. 8.2(c).

In the second example, the envelope bounds were set to $\beta = f + 30$ and $\alpha = f - 30$ where f is the median filter over a 5×5 window. The result is shown in Fig. 8.2(d).

The authors expressed the error in terms of the mean-square error (MSE) criteria. The original image had an MSE of 1912 and this was reduced to 106 by stack filtering. The two envelope approaches further reduced the error to 79 and 55, respectively. This is reflected in the appearance of the filtered images. The second envelope is the most beneficial because it uses the median filter to suppress the most extreme errors. However, this does not affect the remainder of the image, which is accurately restored by the stack filter.

The envelope constraint is very effective in that it basically uses one filter within another. The stack filter is mostly very accurate with a few extreme errors. The median filter always gives a result close to the correct value but causes some local distortions. The envelope constraint combines the best properties of the two filters by principally using the stack filter but limiting its output to be within a set range of the median output. The median guards against extreme errors, but for the vast majority of output samples, it does not influence the final value.

A further aspect of implementation involves designing electronic circuits to carry out the processing. This is covered in the next section.

8.2 Hardware Implementation

The techniques presented in this book thus far may be implemented in software using either a package such as MATLAB or programmed directly in C/C++. Whereas the theory of morphology is documented in terms of set theory and lattices, these operations must be translated into either logical or arithmetic operators when implemented in software or hardware.

The binary imaging work may be implemented in hardware simply by forming a function with inputs consisting of each location in the filter window. An example of this was shown in Figs. 5.7 and 5.9 where the optimal filter for noise removal in a document was designed and implemented.

Application to grayscale processing is more challenging. With the growth in FPGA products the reader may wish to implement some of these methods in hardware. The circuits presented are an illustration of the best approaches and are given to aid understanding of the techniques as well as provide a clue to implementation. The circuits are largely canonical and so are not necessarily the most efficient means of implementation. In practice, hardware optimization and implementation software will reduce the circuits to their minimum form, those requiring a direct route to extremely efficient hardware should see the work of Gasteros who has specialized heavily in this area.[6–9] Other examples of hardware implementation of nonlinear filters are given in the reference section.[10–15]

The general framework for implementing all types of filters presented in this book and based on computational morphology is shown in Fig. 8.3. This framework was described in more general terms in the previous chapter and is placed in the context of hardware here to reinforce the concepts.

The structure presented assumes that the image data is stored as unsigned integers in binary format. The example shows 3-bit data forming a stack of eight threshold levels.

Consistent with computational morphology, the circuit contains three sections:

1. stacking,
2. filtering, and
3. unstacking.

The *stacking* section converts the n binary digits $I_0, I_2, \ldots I_{n-1}$ of a number N, to its L threshold levels $x^0, x^2, \ldots x^{L-1}$, where

$$x^i = 1 \quad \text{for} \quad 0 < i \leq N \quad \text{and} \quad x^i = 0 \quad \text{for} \quad N < i \leq L - 1, \tag{8.2}$$

and $L = 2^n$.

This is produced by a straightforward piece of digital logic design.[16] An example for $n = 3$, $L = 8$ is given in Fig. 8.4. The truth table mapping is given in Fig. 8.4(a). This means that eight functions x^i, for each threshold level i need to be

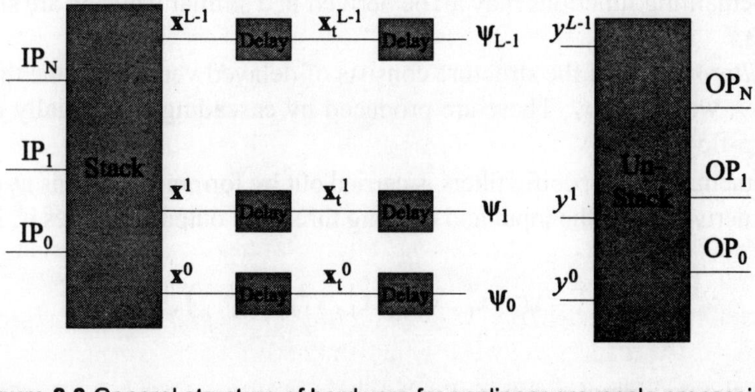

Figure 8.3 General structure of hardware for nonlinear grayscale processing.

I_2	I_1	I_0	x_7	x_6	x_5	x_4	x_3	x_2	x_1	x_0
0	0	0	0	0	0	0	0	0	0	1
0	0	1	0	0	0	0	0	0	1	1
0	1	0	0	0	0	0	0	1	1	1
0	1	1	0	0	0	0	1	1	1	1
1	0	0	0	0	0	1	1	1	1	1
1	0	1	0	0	1	1	1	1	1	1
1	1	0	0	1	1	1	1	1	1	1
1	1	1	1	1	1	1	1	1	1	1

(a)

I_2 \ I_1I_0	00	01	11	10
0	0	0	1	0
1	1	1	1	1

(b)

$$
\begin{aligned}
x_0 &= 1\\
x_1 &= I_2 + I_1 + I_0\\
x_2 &= I_2 + I_1\\
x_3 &= I_2 + I_1 I_0\\
x_4 &= I_2\\
x_5 &= I_2 I_1 + I_2 I_0\\
x_6 &= I_2 I_1\\
x_7 &= I_2 I_1 I_0
\end{aligned}
$$

(c)

Figure 8.4 Design of stacking logic. (a) Truth table for generation of threshold levels x_i from binary data I_i. (b) Example of K map for variable x_3. In the example, x_3 is designed and the function is therefore $x_3 = I_2 + I_1 I_0$. (c) The remaining functions.

derived. The K map for the function corresponding to x^3 has been derived as an example and is shown in Fig. 8.4(b). It can be seen that this corresponds to $x_3 = I_2 + I_1 I_0$. The remaining functions may all be derived in a similar way and are shown in Fig. 8.4(c).

The *filtering* part of the structure consists of delayed versions of the threshold variables x^i, written as x_t^i. These are produced by cascading sequentially clocked D-type flip-flops.

Implementation of specific filters is carried out by forming functions ψ_i of these signals x_t^i derived from the input and creating threshold output variables y^i, i.e.,

$$y^i = \psi_i\left(x_0^0, x_0^1, \ldots x_1^1, x_1^2, \ldots x_{T-1}^{L-1}\right), \tag{8.3}$$

where T is the maximum number of delays.

In the above equation representing the most general filter, the value of the output at every level y^i is a function of all x_t^i, i.e., samples of the input derived from every level i and every time delay t. It was stated in the previous chapter that Dougherty[17] has shown that any filter, linear or nonlinear, may be represented in terms of computational morphology and hence may be placed in this form.

In practice, special cases of filters such as stack filters and grayscale morphology result in restricted forms of the functions ψ_i.

The *unstacking* operation consists of digital logic that converts the maximum value of i for which $y^i = 1$ to a binary number. This is a matter of straightforward (if tedious) logic design. The interesting part of the process lies in the functions ψ_i which link y^i and x_t^i.

8.3 Stack Filter

The simplest case of nonlinear filtering in this context is the *stack filter*. Figure 8.5 shows the structure of a stack filter with $T = 3$ and $L = 4$. The general model is simplified to

$$y^i = \psi\left(x_0^i, x_1^i, \ldots x_{T-1}^i\right) \tag{8.4}$$

The value of y^i is determined by a combinatorial binary function of the time-delayed versions of x^i. Note that y^i (the output at level i) is only dependent on the inputs x_t^i, i.e., those derived from the same threshold level i. Also, the binary function ψ is the same for all levels: $\psi_i = \psi$. Many increasing filters may be represented via stack filters including morphological operations with flat structuring elements and the median and weighted median. Some linear FIR filters with positive coefficients may also be represented as a stack filter.

A specific case of the stack filter is the morphological operation of a *three-point erosion* of the signal X by a flat structuring element B to produce an output signal Y, i.e.,

$$Y = X \Theta B \tag{8.5}$$

where Θ is the erosion operator and B is a three-point flat structuring element. In terms of logic, this reduces to a three-input Boolean AND operation applied to the individual elements of the signals, i.e.,

$$y^i = x_0^i \cdot x_1^i \cdot x_2^i \tag{8.6}$$

The output signal therefore consists of an AND function of the time-delayed inputs at the same threshold level. This is shown in Fig. 8.6.

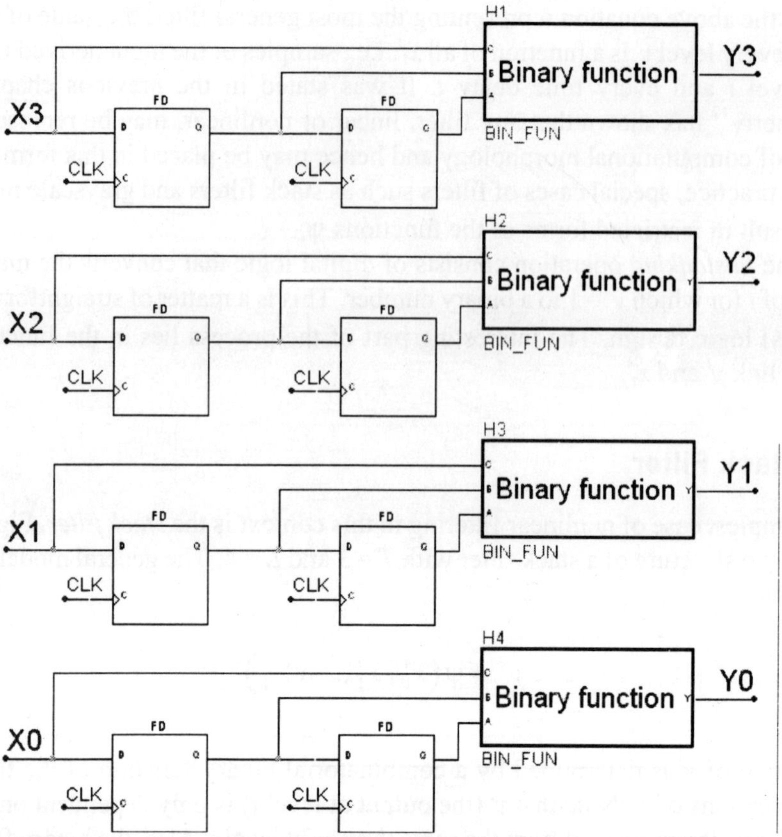

Figure 8.5 Stack filter implementation. The output value at each stack level is a binary function of the thresholded inputs at the same level.

The stack filter may also be used to implement some rank-order and weighted-order filters. In fact, the maximum operator is precisely the same as the erosion operator described above. From the description of stacking in the previous chapter, the median filter may be implemented via a stack filter using the majority function. For three inputs this may be written as

$$y^i = x_0^i \cdot x_1^i + x_1^i \cdot x_2^i + x_0^i \cdot x_2^i. \tag{8.7}$$

This is shown in Fig. 8.7.

Stack filters consist of levels of independently computed values. This has the advantage that each level may be implemented in parallel without reference to the others. This means that what began as a gray level *sorting* operation, i.e.,

$$y = \text{median}(x_0, x_1, x_2), \tag{8.8}$$

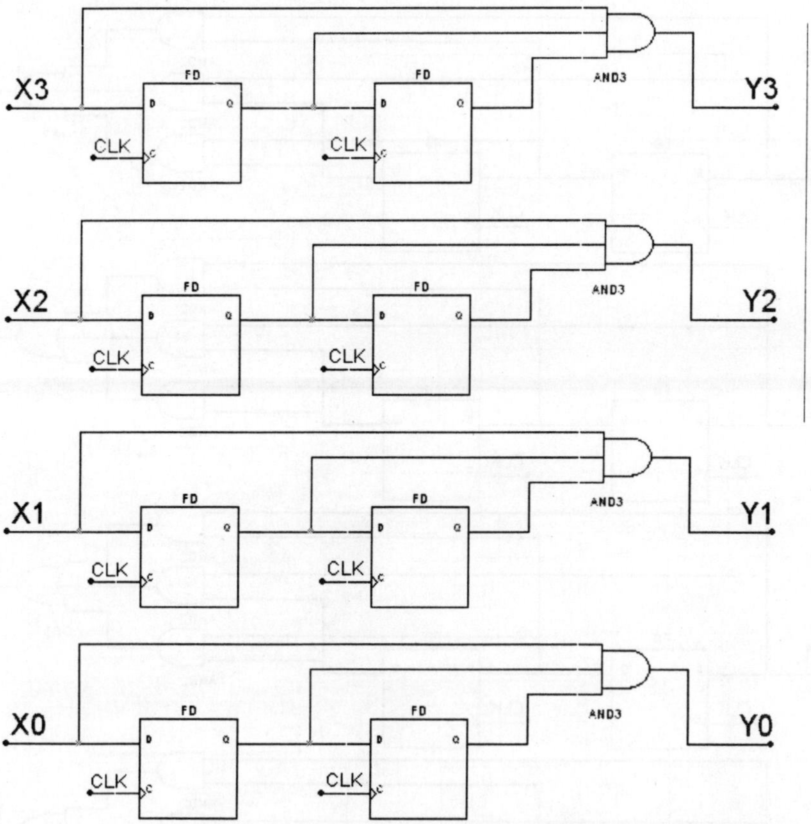

Figure 8.6 Erosion by a three-point flat structuring element implemented via a stack filter.

has been replaced by a simple *logical* operation. The circuit shown in Fig. 8.7 produces an output in the propagation time of a simple logic gate with no sorting required.

It may also be implemented as a simple *counting* operation, i.e.,

$$y^i = \begin{cases} 1 & \text{if} \quad x_0^i + x_1^i + x_2^i \geq 2 \\ 0 & \text{otherwise} \end{cases} \tag{8.9}$$

For larger median filters, the counting operation may be preferable since the implementation in terms of logic does not scale well. Rather than duplicating the digital logic for every threshold level, other strategies for computing stack filters have been adopted. By definition, the output values of a stack filter are $y^i = 1$ for $i \leq k$ and $y^i = 0$ for $i > k$. This means that the only level of interest is k, the highest threshold level for which the output is 1 (or the top of the signal). Various approaches have been developed to locate this level.[18] One approach is a divide-and-conquer strategy where the value of the middle threshold level $y^{L/2}$ is computed. If $y^{L/2} = 1$, then k

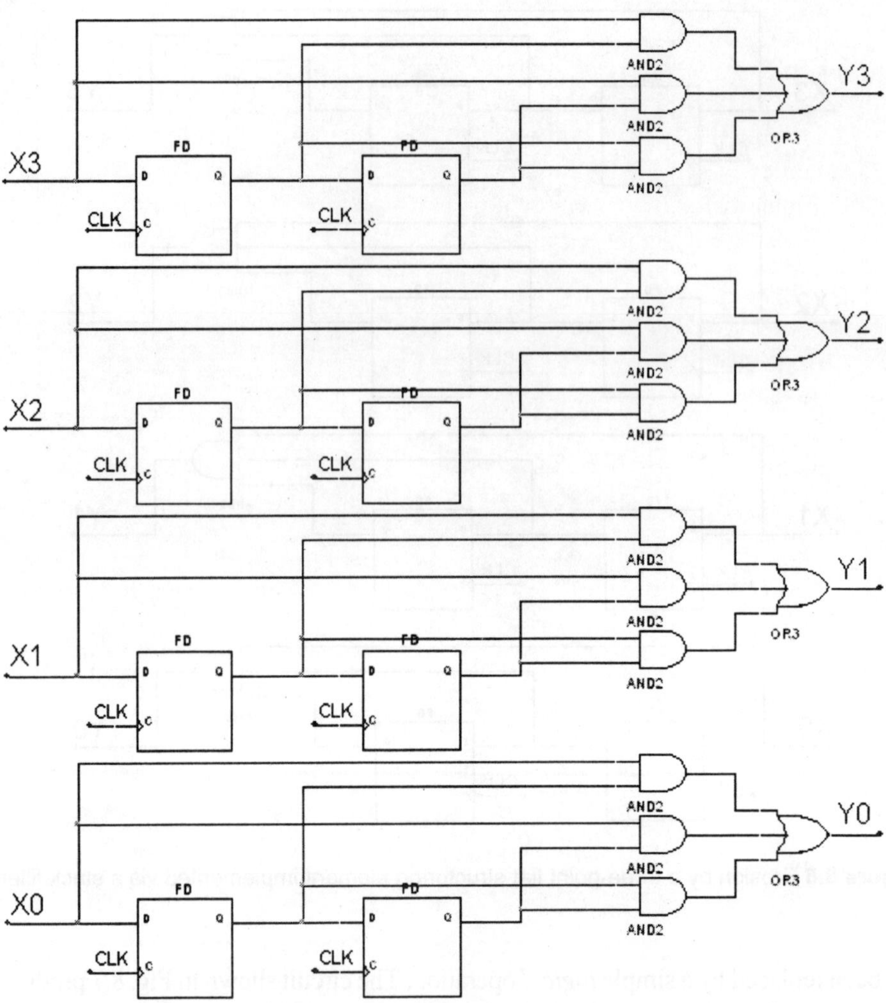

Figure 8.7 Three-point running median filter implemented via a stack filter.

lies in the top half of the dynamic range of the output, i.e., $L/2 \leq k \leq L-1$. Otherwise it lies in the bottom half, so $0 \leq k \leq L/2$. Next the value in the middle of the top half $y^{3L/4}$ is computed and the process continues until the precise value of k is located.

An elegant way of implementing this strategy is using a *bit-serial approach*. The gray level value of the output signal y is determined one bit at a time starting with the most significant bit. The value of each bit indicates if the output lies in the top or bottom half of the range.

A further observation that can result in a drastic reduction in computation time is that the grayscale output y must always be the same as one of the time-delayed (unthresholded) input values x_t, i.e., given that

$$y^i = \psi\left(x_0^i, x_1^i, \ldots x_{T-1}^i\right), \tag{8.10}$$

then $y = x_j$ where $x_j \in \left(x_0, x_1, \ldots x_{T-1}\right)$.

This means that a stack filter need only compute as many threshold levels as there are unique values in the input window. So a stack filter with T input variables may be implemented by working with, at most, T threshold levels rather than compute the full dynamic range, typically 256 for 8-bit data. This method is called "range compression" by Lin et al.[19] Range compression is illustrated in Fig. 8.8.

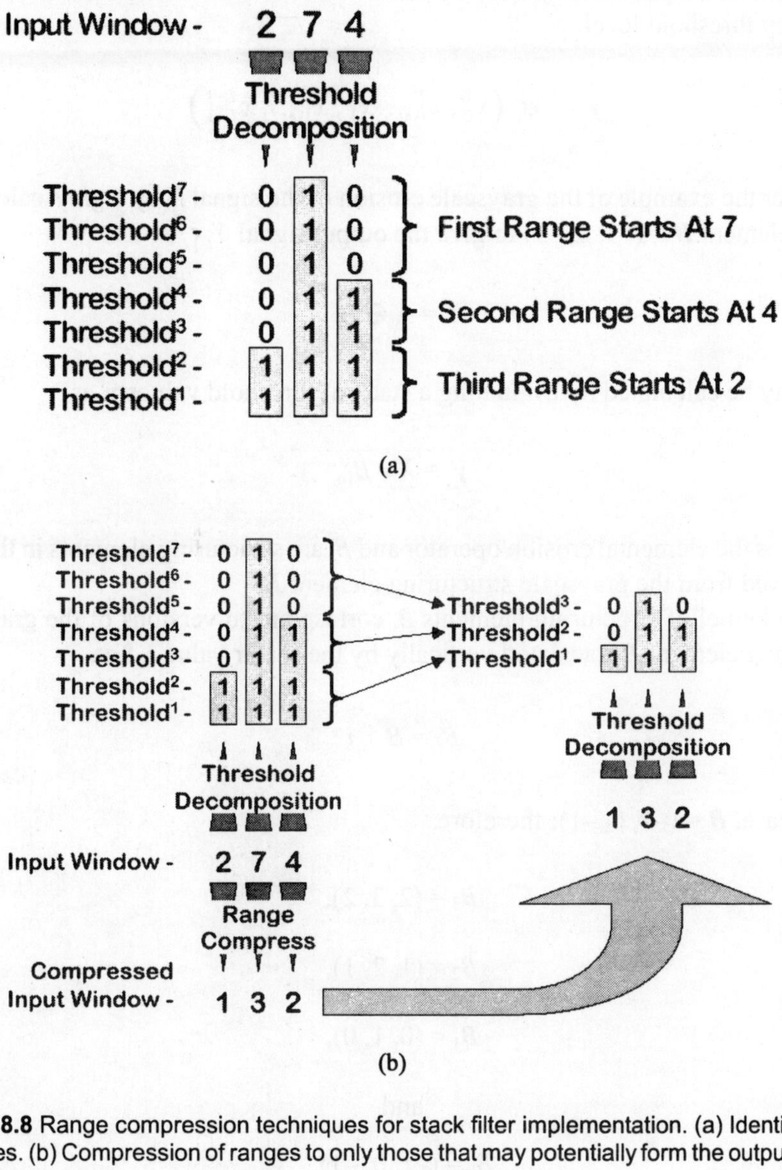

Figure 8.8 Range compression techniques for stack filter implementation. (a) Identification of ranges. (b) Compression of ranges to only those that may potentially form the output value.

Various algorithms for efficient stack filter implementation especially in FPGA hardware are presented in Woolfries.[20]

8.4 Grayscale Morphology

Grayscale morphology may be implemented as a special case of the more general computational morphology (CM). The overall CM structure is simplified so that the processing function ψ_i is the same at every threshold level, i.e., $\psi_i = \psi$. This remains more general than the stack filter configuration, however, since the individual threshold output values y^i may be formed as a function of input values derived from any threshold level.

$$y^i = \psi_i\left(x_0^0, x_0^1, \ldots x_1^1, x_1^2, \ldots x_{T-1}^{L-1}\right) \tag{8.11}$$

Consider the example of the grayscale erosion of the signal X by a grayscale structuring element $\boldsymbol{B} = (-1, 0, -1)$ to give the output signal Y:

$$Y = X \Theta \boldsymbol{B} \tag{8.12}$$

This may be calculated by evaluating a stack of threshold values y^i as

$$y^i = X\, \boldsymbol{e}\, B_i, \tag{8.13}$$

where $\boldsymbol{e}$ is the elemental erosion operator and B_i are structuring elements in the kernel derived from the grayscale structuring element $\boldsymbol{B}$.

The kernel of structuring elements B_i correspond to versions of the grayscale structuring element $\boldsymbol{B}$ translated vertically by the scalar value i, i.e.,

$$B_i = \boldsymbol{B} + i \tag{8.14}$$

In this case, $\boldsymbol{B} = (-1, 0, -1)$; therefore,

$$B_3 = (2, 3, 2),$$

$$B_2 = (1, 2, 1),$$

$$B_1 = (0, 1, 0),$$

and

$$B_0 = (-1, 0, -1).$$

The elemental erosion tests if the corresponding structuring element B_i fits beneath the input signal. While in computational morphology theory the input signal is not thresholded, in practice this test is put into the context of a series of threshold inputs. This is explained by the following function

$$y^i = x_0^{i-1} \cdot x_1^i \cdot x_2^{i-1}. \tag{8.15}$$

It can be seen that the output at each threshold level y^i is formed by taking an AND function of the three inputs (x_0^{i-1}, x_1^{i}, x_2^{i-1}). These correspond to the (temporally) central input signal x_1^i (from the same level i) and the two on either side x_0^{i-1} and x_2^{i-1} (taken from the next level down, i–1). Provided that all three of these signals are 1, then the structuring element B_i fits below the input signal x and hence the output at that threshold level $y^i = 1$.

For consistency it is usually assumed that inputs $x_i = 1$ for $i < 0$, hence the outputs are defined as

$$\begin{aligned} y^3 &= x_0^2 \cdot x_1^3 \cdot x_2^2 \\ y^2 &= x_0^1 \cdot x_1^2 \cdot x_2^1 \\ y^1 &= x_0^0 \cdot x_1^1 \cdot x_2^0 \\ y^0 &= x_1^0. \end{aligned} \tag{8.16}$$

The circuit to implement this grayscale morphological erosion is shown in Fig. 8.9.

8.5 Computational Morphology and Aperture Filters

A general computational morphology filter would be capable of implementing any function of any complexity, linear or nonlinear. However, the function used to create each output level would require access to all levels of the thresholded input. It would therefore be very complex.

It would also be very difficult to design, since the unconstrained filter requires that every possible combination of inputs is seen a sufficient number of times to estimate the output conditional probabilities.

Design of such filters is normally complex and consists of determining the content of the kernel of structuring elements. This translates into estimating the functions ψ_i in Eq. 8.3. It is essential that the functions are ordered in such a way that

$$\psi_i \geq \psi_{i+1} \text{ for all } i \text{ and for all combinations of inputs.} \tag{8.17}$$

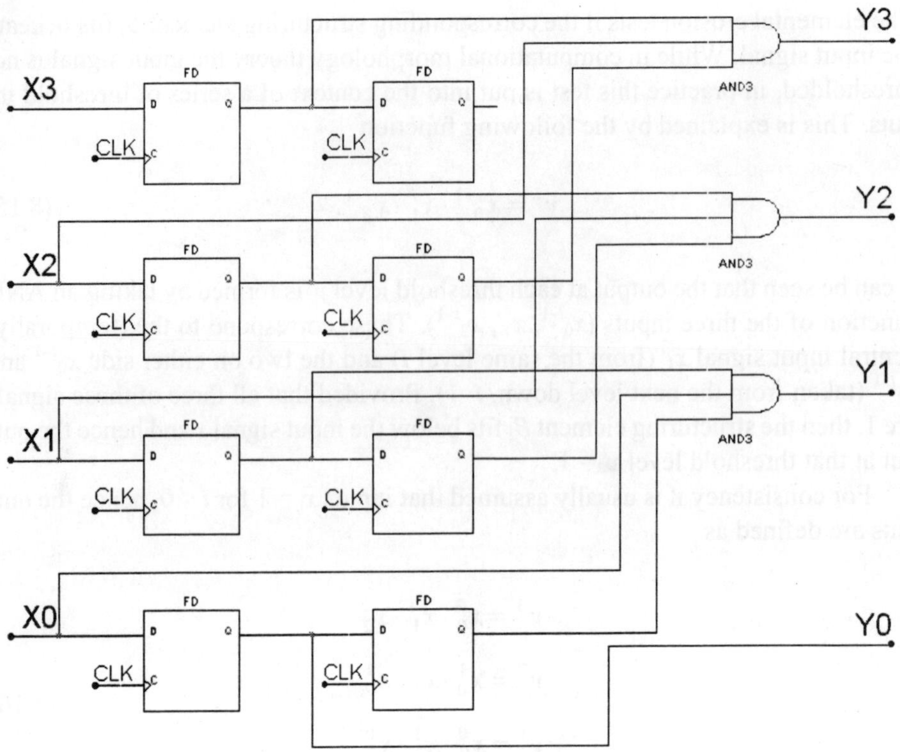

Figure 8.9 Grayscale erosion by structuring element (–1, 0, –1).

This is important in order to preserve the stacking property of the output signals y^i. In stack filters this ordering is guaranteed by using only positive Boolean functions. For computational morphology, Dougherty has named this property *consistency*.

In practice, computational morphology is often too general for most applications and special cases of it are adopted. One practical method is that of aperture filters described in the previous chapter. Aperture filters are very similar to those based on computational morphology, but with a much reduced dynamic range. This is achieved by subtracting a signal similar to a moving average corresponding to the aperture placement function. This signal is added back onto the aperture filter output following filtering.

An aperture placement signal P is calculated as a running function ρ of X:

$$P = \rho(X) \tag{8.18}$$

This signal P is subtracted from the input signal X giving the aperture filter input. A representation of this process is given in Fig. 8.10.

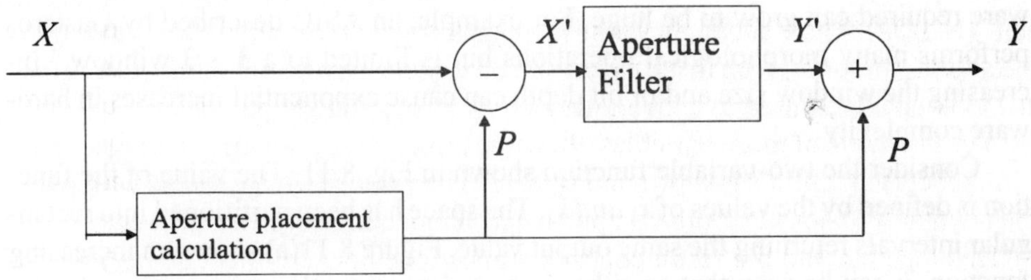

Figure 8.10 Aperture filter placement.

$$X' = X - P \tag{8.19}$$

The aperture filter is applied to X' to give the output I':

$$Y' = \psi(X') \tag{8.20}$$

The offset signal P is added back to the aperture output to get the overall output signal Y:

$$Y = Y' + P \tag{8.21}$$

$$Y = \rho(X) + \psi[X - \rho(X)]$$

The reduced dynamic range of aperture filters makes them much easier to design. Unlike the earlier filters described in this chapter, the amplitude values are both positive and negative but there are no conceptual problems with this extension. Further constraints may be applied within the aperture filter if necessary.

8.6 Efficient Architecture for Computational Morphology and Aperture Filters

As relatively new concepts, hardware implementations of computational morphology and aperture filters are still emerging. While in theory the translation from algorithm to logic is straightforward for many practical problems, the complexity can increase very rapidly. One recent novel approach to implementation involving bit vector architecture was proposed by Handley.[21] Computational morphology partitions all possible combinations of inputs into a series of intervals with a specific output associated with that interval. Its implementation reduces to a search problem in order to determine in which interval a set of windowed observations lie. Comparator-based architectures perform this task in parallel[22,23] and the amount of hardware required can grow to be huge. For example, an ASIC described by Gasteros performs many morphological operations but is limited to a 3×3 window.[9] Increasing the window size and/or bit depth can cause exponential increases in hardware complexity.

Consider the two-variable function shown in Fig. 8.11. The value of the function is defined by the values of x_1 *and* x_2. The space has been partitioned into rectangular intervals returning the same output value. Figure 8.11(a) shows an increasing function, it can be seen that as either x_1 or x_2 increases, the output value also increases. Figure 8.11(b) shows a nonincreasing function, since no such relationship holds.

The problem here is to determine the interval given x_1 and x_2. Notice that the same output may result from separate disjoint intervals. The approach in compara-

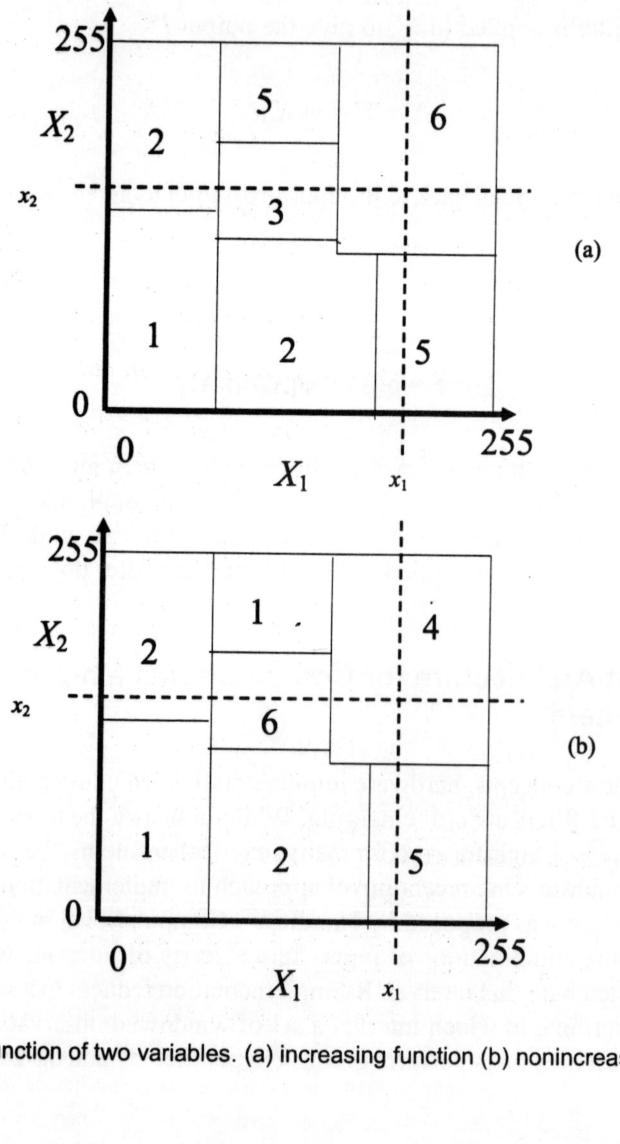

Figure 8.11 Function of two variables. (a) increasing function (b) nonincreasing function.

tor-based solutions is to construct a series of parallel detectors, one for each output value. The amount of hardware required can be massive. It can be reduced by sequential testing, but this can be slow and result in non-deterministic processing times.

An alternative approach is the bit-vector architecture. The concept is illustrated by Handley through a simple example which is repeated here and in Fig. 8.12. The function space is partitioned into a series of intervals taking the shape of disjoint hyper-rectangles. In the two-variable case shown here, these correspond to simple rectangles. Each interval has the same value of output associated with it. The numbers shown on each interval are labels (rather than output values). The intervals with similar shading have the same output, therefore 2 and 5 have the same output as do 1 and 6, and 0, 3, 4, and 7. The area outside of these intervals represents "no operation." Any single value of each variable will intersect a number of the intervals. For example, in Fig. 8.12, the value of the variable x_1 shown intersects intervals 2, 3, and 6. This is coded in a bit vector as (0, 0, 1, 1, 0, 0, 1, 0). Each interval 0 through 7 is represented (left to right) by a single bit. Each bit is **1** if the interval has been intersected and **0** if it has not. Similarly, the value of the variable x_2 shown intersects intervals 1 and 3. This is coded in a bit vector as (0, 1, 0, 1, 0, 0, 0, 0). To determine the interval at location (x_1, x_2) the two bit vectors are simply ANDed together to compute their intersection, which in this case is (0, 0, 0, 1, 0, 0, 0, 0), or interval 3. It may be that the intersection forms the empty set in which case this is a "No opera-

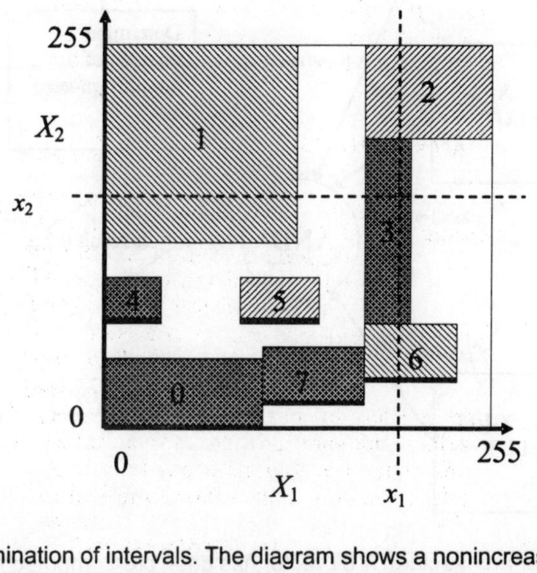

Figure 8.12 Determination of intervals. The diagram shows a nonincreasing function of two variables x_1 and x_2. The space is partitioned into a series of hyper-rectangles each returning a single value. A look-up table of bit vectors, indicating which intervals are intersected, is pre-computed for each value of x_1 and x_2. In the example above, the first variable x_1 returns a bit vector of (0,0,1,1,0,0,1,0) representing intervals 2, 3, and 6. The second variable x_2 returns a bit vector of (0,1,0,1,0,0,0,0) representing intervals 1 and 3. The intersection of these bit vectors is (0,0,0,1,0,0,0,0) identifying the correct interval as 3.

tion" (NO OP). There should never be more than one nonzero bit in the intersection, and a formal proof is given in the paper. The bit vectors must be pre-computed. A further look-up take is used to convert the interval label to the output value.

For functions of more variables, the same principle applies and an architecture for nonincreasing filters is shown in Fig. 8.13. The value of each variable within the filter window maps to a pre-computed bit vector. All of these bit vectors are ANDed together and the result contains at most one nonzero bit. The position of this bit identifies the label of interval required and this label is then converted to the output value via a look-up table. If there are no nonzero bits, then the result is no operation. A slightly simpler version of the architecture exists for increasing filters.

In practice, computational morphology is too general for many applications and aperture filters are used instead. Aperture filters operate over a smaller set of input values by clipping the input range into the filter window. They partition the input values into intervals and return an output for each interval. This is therefore ideally suited to implementation in a bit-vector architecture.

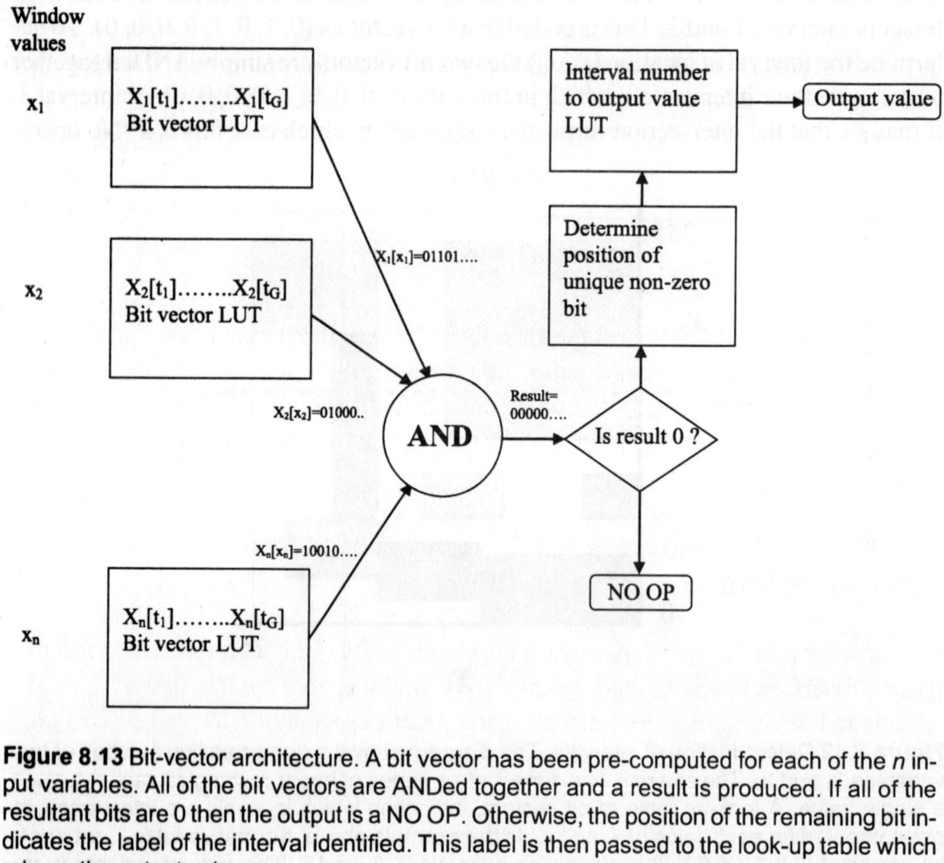

Figure 8.13 Bit-vector architecture. A bit vector has been pre-computed for each of the n input variables. All of the bit vectors are ANDed together and a result is produced. If all of the resultant bits are 0 then the output is a NO OP. Otherwise, the position of the remaining bit indicates the label of the interval identified. This label is then passed to the look-up table which returns the output value.

8.7 Summary

This chapter has considered a range of issues that arise in the practical implementation of the techniques described earlier in the book. The first part concerned envelope filters, which are a useful technique for reducing the rare gross errors that can occur in certain types of filters. The remainder of the chapter has been concerned with hardware implementation. No matter how well the methods work, they will not be widely adopted if there are serious problems with implementation.

While the techniques appear to map to hardware in a straightforward way, they do not in general scale well. Small increases in window size and bit depth can cause very rapid increases in the hardware required making it impractical in some situations. A number of smarter methods for implementation including bit-serial approaches and bit-vector architecture have been presented that result in efficient implementations.

References

1 Vo Ba-Ngu Vo, Antonio, C., "Continuous-time envelope constrained filter design with input uncertainty," *ICASSP98*, **3**, 1289–1293 (1998).

2 K. L. Teo, A. Cantoni, and X. G. Lin, "A new approach to the optimization of envelope-constrained filters with uncertain input," *IEEE Transactions on Signal Processing*, **42**(2), 426–429 (1994).

3 W. X. Zheng, A. Cantoni, and K. L. Teo, "Robust design of envelope-constrained filters in the presence of input uncertainty," *IEEE Transactions on Signal Processing*, **44**(8), 1872–1877 (1996).

4 C. H. Tseng, K. L. Teo, A. Cantoni, and Z. Zang, "Envelope-constrained filters: adaptive algorithms," *IEEE Transactions on Signal Processing*, **48**(6), 1597–1608 (2000).

5 M. Brun, R. Hirata Jr., J. Barrera, and E. R. Dougherty, "Nonlinear filter design using envelopes," *J. Math. Imaging Vis.*, **21**(1), 81–97, 2004.

6 A. Gasteratos, I. Andreadis, and P. Tsalides, "Improvement of the majority gate algorithm for gray-scale dilation/erosion," *Electronics Letters,* **32**(9), 806–807 (1996).

7 A. Gasteratos, I. Andreadis, and P. Tsalides, "Extension and very large scale integration implementation of the majority-gate algorithm for gray-scale morphological operations," *Opt. Eng.*, **36**(3), 857–861 (1997).

8 A. Gasteratos, I. Andreadis, and P. Tsalides, "Realisation of soft morphological filters," *IEE Proceedings – Circuits Devices and Systems*, **145**(3), 201–206 (1998).

9 A. Gasteratos and I. Andreadis, "Non-linear image processing in hardware," *Pattern Recognition*, **33**(6), 1013–1021 (2000).

10 I. Diamantaras and S. Y. Kung, "A linear systolic array for real-time morphological image processing," *J. VLSI Signal. Proc.*, **17**(1), 43–55 (1997).

11 S. J. Ko, A. Morales, and K. H. Lee, "A fast implementation algorithm and a bit serial realization method for grayscale morphological opening and closing," *IEEE Transactions on Signal Processing*, **43**(12), 3058–3061 (1995).

12 L. Lucke and C. Chakrabatri, "A digit-serial architecture for gray-scale morphological filtering," *IEEE Transactions on Image Processing*, **4**(3), 387–391 (1995).

13 L. Abbott, R. M. Haralick, and X. Zhuang, "Pipeline architectures for morphologic image analysis," *Machine Vision and Applications*, **1**(1), 23–40 (1988).

14 I. Diamantaras, K. H. Zimerman, and S. Y. Kung, "Integrated fast implementation of mathematical morphology operations in image processing," *IEEE International Symposium on Circuits and Systems*, New Orleans, 1442–1445 (1990).

15 D. S. Bloomberg, "Implementation efficiency of binary morphology," *ISMM 2002*, Sydney, Australia (2002).

16 C. H. Roth, *Fundamentals of Logic Design*, 4th ed., Brooks Cole, New York (1995).

17 E. R. Dougherty and D. Sinha, "Computational mathematical morphology," *Signal Processing*, **38**, 21–29 (1994).

18 K. Chen, "Bit-serial realisation of a class of non linear filters based on positive Boolean functions," *IEEE Trans. Acoustics, Speech and Signal Processing*, **ASSP-36**(6), 785–794 (1989).

19 L. Lin, G. B. Adams, and E. J. Coyle, "Input Compressed and Efficient Algorithms and Architectures for Stack Filters," *Proc IEEE Winter Workshop on Nonlinear Digital Signal Processing*, Tampere, Finland, 5.1–5.4 (1993).

20 N. Woolfries, *Efficient Hardware Implementation of Stack Filters Using FPGAs*, MPhil thesis, University of Strathclyde, Glasgow, Scotland (2002).

21 J. C. Handley, "Bit vector architecture for computational mathematical morphology," *IEEE Transactions on Image Processing*, **12**(2), 153–158 (2003).

22 E. R. Dougherty and D. Sinha, "Computational gray-scale mathematical morphology on lattices (a comparator-based image algebra)—Part I: Architecture," *Real-Time Imaging*, **1**(1), 69–85 (1995).

23 E. R. Dougherty and D. Sinha, "Computational gray-scale mathematical morphology on lattices (a comparator-based image algebra)—Part II: Image operators," *Real-Time Imaging*, **1**(4), 283–295 (1995).

Chapter 9

Case Study: Noise Removal from Astronomical Images

The previous chapters have included examples of filtered images and a discussion of the implementation of morphological and logic-based filters. This chapter presents a case study showing how a morphological filter may be designed for a specific type of noise in images, namely astronomical images.

Imaging instrumentation is widely used in space-based astronomy and solar physics where it has the potential to produce excellent pictures. However, these are frequently degraded by bursts of cosmic ray ions that saturate the charge-coupled devices (CCDs) and produce an overlaid speckle. This is a source of frustration to observers and can obscure vital detail. In this chapter it will be shown that the speckle may be removed from the image using a type of nonlinear filter known as a soft morphological filter.

Soft morphological filters comprise a branch of nonlinear image processing that is particularly effective for noise removal. They originate from the field of mathematical morphology but their operations are less harsh since the structuring elements used are designed to have "soft" boundaries. The implementation of such filters makes extensive use of rank-ordering operations.

The chapter will describe how a training set may be created for the images and how the optimal filters may be derived using genetic algorithms. The results of processing the images with the optimal filters will be presented. Finally, experiences of implementing the filters in programmable hardware will be given.

9.1 CCD Noise in Astronomical and Solar Images

CCDs are used in many space-based astronomy and solar physics imaging instruments, such as the Hubble Space Telescope's wide-field and planetary cameras, the Solar and Heliospheric Observatory (SOHO), the Extreme Ultraviolet Imaging Telescope (EIT), and the Large Angle Solar Coronagraph (LASCO[1]).

Figure 9.1 Example of an image of the sun's corona. This image is very clean with no CCD speckle.

In the space environment, CCDs are bombarded by cosmic ray ions (CRs) that register as counts in the CCD. Therefore, for space-based use, CCDs are shielded and radiation-hardened to minimize the permanent damage. In particularly high-radiation environments, radiation hits can significantly degrade the quality and usefulness of the data. Even for ground-based observations with long integration times, CR hits can be problematic.

Figure 9.1 shows an example of an image taken with LASCO onboard the SOHO. This image is very clean with no evidence of CR speckle. Conversely, the image shown in Fig. 9.2 has suffered significant particle hits. This was caused by high energy particles accelerating close to the sun and impacting the CCD leading to the characteristic "snow" that in some cases almost completely whites out the image.[2]

The effects of this speckle noise may be significantly reduced by the application of a correctly designed soft morphological filter.

Figure 9.2 The problem of CCD overload is shown. Cosmic rays have hit the CCD and have caused the image to be heavily distorted by noise. The removal of this noise is the subject of this case study.

9.2 Soft Morphological Filters

It has been shown in earlier chapters that the amount of training data required for the accurate design of filters grows very rapidly with increasing filter size. However, in order to solve many practical imaging problems, filters at least as large as 5 × 5 pixels are required. Unless these filters are constrained in some way, the search space will become impossibly large. For example, recall that even a 5 × 5 stack filter has 2^{25} (i.e., more than 33 million) input combinations. This means that each of these would have to be observed several times in order to produce an accurate estimate of their conditional output probability.

Therefore, it is necessary to constrain the filter to reduce this search space substantially. Many approaches to constraining the filters result in methods requiring the rank ordering of the data.

Soft morphological filters comprise a branch of nonlinear image processing particularly effective for noise removal. The soft morphological filter was first introduced by Koskinen[3] in 1994. Design techniques for these filters based on genetic algorithms were developed by Marshall and Harvey[4,5,6,7] and applications include spatio-temporal filters for film archive restoration.[8]

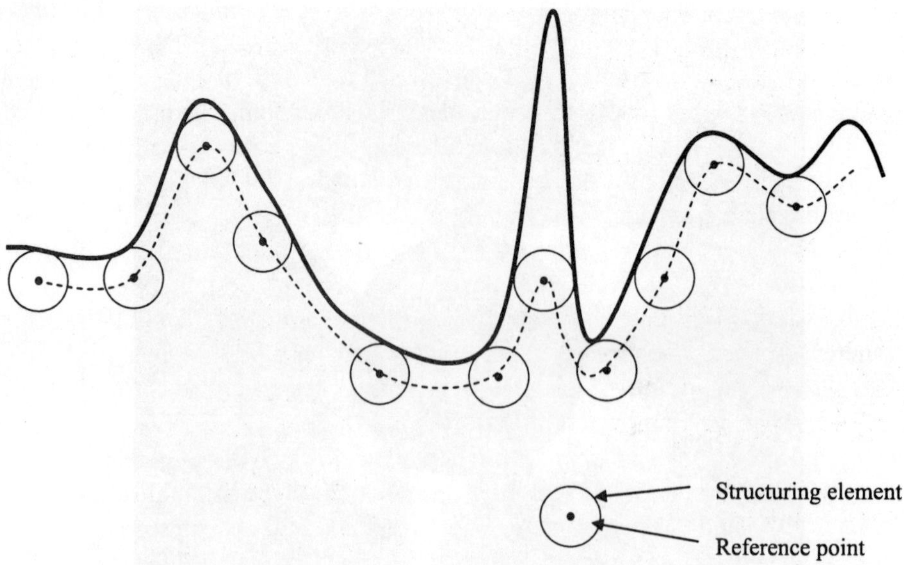

Figure 9.3 Standard morphological erosion by a structuring element. The dotted line indicates the signal after filtering (i.e. the eroded signal).

Soft morphology is slightly more subtle than standard grayscale morphology. First, let us remind ourselves of the operation of grayscale morphology. A brief non-mathematical overview of standard grayscale morphology is provided here followed by an equivalent description of soft morphology in order to allow the reader to distinguish between them. Readers requiring a more mathematically rigorous explanation should consult the references.

Figure 9.3 shows a sketch of a standard grayscale erosion of a 1D signal, using a circular structuring (SE). The SE is "pushed up" from below so that it just touches the signal. The SE then slides along the signal, moving up and down while maintaining contact from below. The filtered signal is given by the path mapped out by the reference point of the SE (which is shown as a dotted line in Fig. 9.3). In this case, the reference point is located at the center of the SE. Note that the whole of the SE remains below the signal at all times. In general, erosion lowers the overall level of the signal, and peaks that are too narrow to contain the SE are removed. By contrast, valleys remain unchanged.

Similarly, the standard 1D grayscale dilation of the same signal by the same SE is given in Fig. 9.4. In this case, the SE is lowered onto the surface of the signal and slides along it, moving up and down as necessary, while remaining in contact from above. Again, the filtered signal is given by the path mapped out by the reference point of the SE and the whole of the SE remains above the signal at all times. In the case of dilation, valleys that are too narrow to contain the SE are filled in, whereas the peaks remain unchanged.

From the above examples, it has been emphasized that the *whole* of the structuring element must remain below the surface for erosion or above the surface for

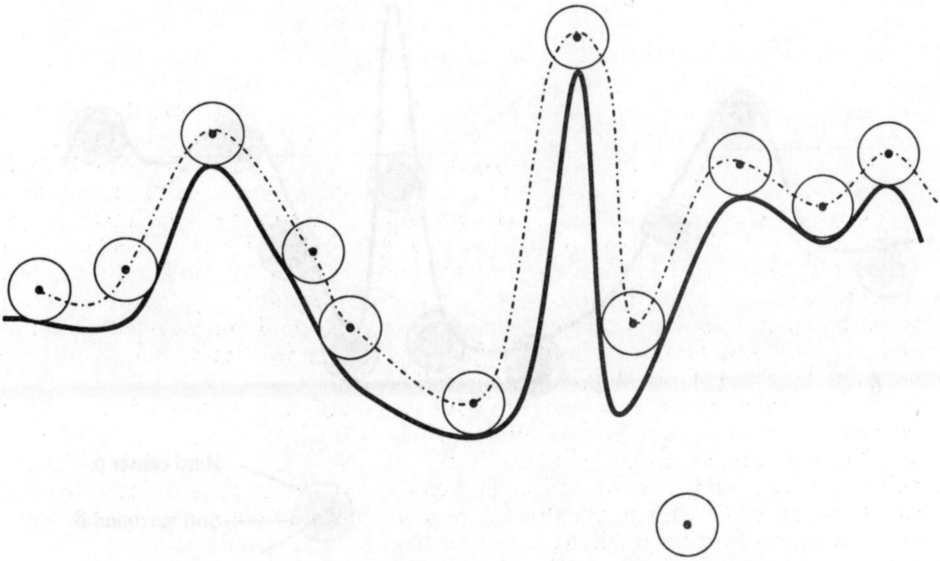

Figure 9.4 Standard morphological dilation by a structuring element.

dilation at all times. However, in soft morphology this constraint is relaxed so that only a certain percentage of the SE is forced to lie below the surface for erosion, or above it for dilation. In fact, in soft morphology the structuring element is partitioned into two regions, a *hard center* α and a *soft surround* β. The hard center behaves in a similar way to the structuring element in standard morphology. That is, for soft erosion the *whole* of the hard center must be "beneath" the signal surface. On the other hand, only a proportion of the soft surround must lie beneath the signal surface. The amount of the soft surround that is forced to lie beneath the surface is controlled by a value *r*, known as the repetition parameter. Conversely, for soft dilation a proportion of the soft surround must remain above the signal.

Soft morphological filtering is therefore a function of three parameters: α, β and *r*. The first two, α and β, specify pixels within the structuring element, and *r* is a scalar quantity that defines what proportion of β must lie either below or above the surface for soft erosion and soft dilation, respectively. By adjusting the three parameters, a more subtle filtering effect is produced. Examples of soft morphological erosion and dilation are given in Fig. 9.5(a) and Fig. 9.5(b), respectively.

Notice that this is a less harsh process. By careful design of the structuring elements, soft morphological filters can by used to remove different types of noise from images while leaving the important structures intact. The design is carried out by a training process using representative examples. In this way, the filter models the inverse process and produces an optimum mapping from noisy to restored image.

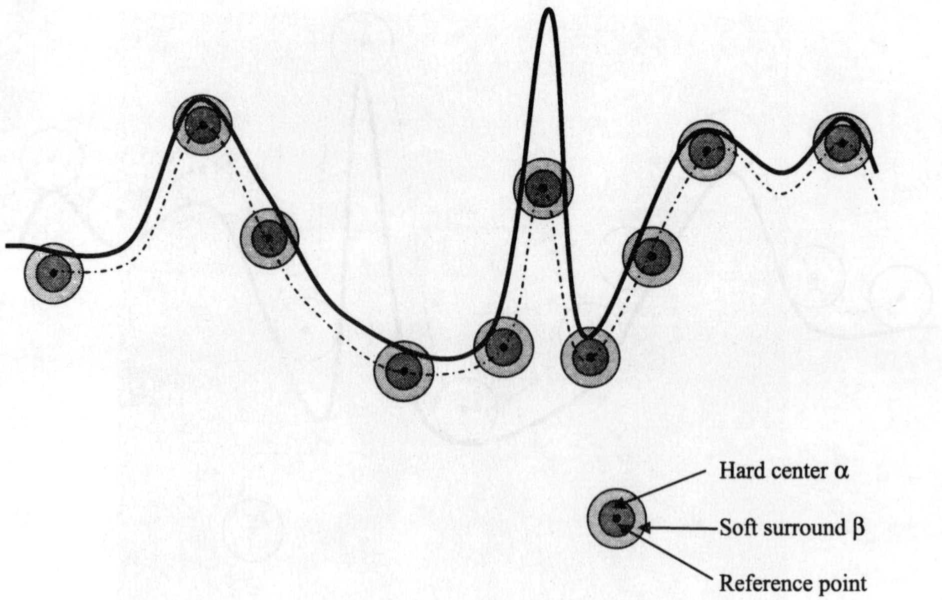

Figure 9.5(a) Soft morphological erosion by a structuring element.

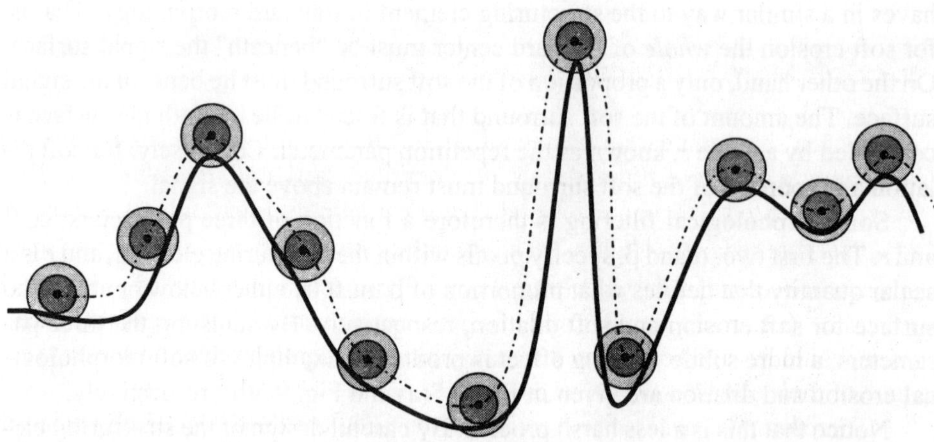

Figure 9.5(b) Soft morphological dilation by a structuring element.

9.3 Results

The soft morphological filter used in this work was designed with a genetic algorithm.[9] The nonlinearities in the filter make it difficult to produce a deterministically designed optimum solution. Instead, an iterative search approach is used that tests a number of different solutions. It combines and modifies these solutions until no further improvement is possible.

The training and application of the soft morphological filter was performed in three steps:

1. A training set was created that mimicked the effect of the disturbance in the real data.
2. Training was performed using a genetic algorithm (GA) and the improvement using training data was confirmed.
3. The resulting filter was applied to the real data and the improvement was observed.

9.3.1 Creation of a training set

The first step in removing the noise from any image is to understand the nature of the disturbance. In this case, the distortion was caused by cosmic rays hitting the CCD in the SOHO telescope and causing the cells to overload, producing an image that suffers from extreme "white out." In many cases, the obstruction was severe enough to render the data worthless in its current form. It was vital to ensure that the noise model used in training was appropriate to that affecting the real images. A poorly chosen noise model will lead to a poorly trained filter.

In order to create a filter to remove this noise, a method of filter training was required. A genetic algorithm was used to determine the optimum filter applied to a training set of representative ideal and noisy image pairs. The noisy image was filtered and the output compared with the ideal image. The GA was used to adjust the filter parameters iteratively in order to make the output as close as possible to the clean image.

The representative training set was created using ideal images, and adding noise from the real images to create a set of noisy images. For this example, clean images from the SOHO telescope were taken from sohowww.nascom.nasa.gov. These images were cropped from 1024 × 1024 pixels to 150 × 150 pixels to reduce training time.

The iterative nature of the filter design can result in large processing times. It is important that the balance between training set size and the search space of the filter, as discussed in Chapter 4, is maintained. It was found from experience that using ten cropped images produces well-trained filters. Once the images were resized, white patches of speckle noise, similar to those seen in Fig. 9.2, were added to create the corresponding noisy images. This was done manually through a cut and paste operation. Together the two sets of images formed the training data.

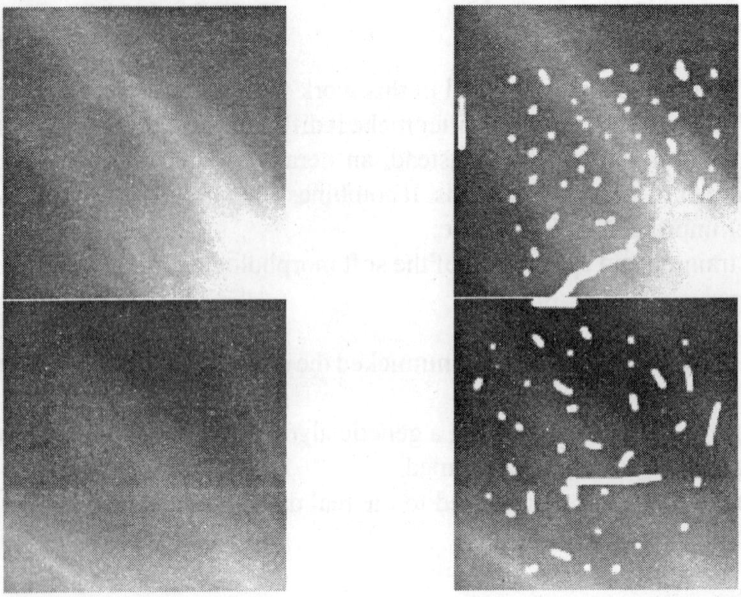

Figure 9.6 Examples of two pairs of training images. The images on the left are clean data. The images on the right are the same pictures with representative noise added. In total, ten of these sets were used in the training process.

Figure 9.6 shows examples of training images used. The images on the left are the original clean ideal images and the ones on the right are the corresponding noisy images created by adding patches of noise manually.

9.3.2 Training

Having created the training set, the next step was to carry out the training process. This was performed using a combination of Matlab and C++ functions. Matlab functions were used to make the overall procedure more scriptable. C++ functions were used for the more computationally intensive parts of the GA to improve the performance of the system.

The genetic algorithms operate by modeling the evolutionary processes found in nature. The filter parameters (α, β, and r) were encoded into a binary string. A fixed number of bits were used to represent the values within the hard center, the soft surround, and the repetition parameter. Collectively these are known as a chromosome.

At the beginning of the training procedure, a population consisting of thirty of these chromosomes was created using a pseudo-random number generator. Each chromosome was translated to a different filter that was applied to the noisy image and its performance was evaluated. The genetic algorithm then proceeded to model

the evolutionary process by the application of *crossover* and *mutation*. Two chromosomes were chosen at random as "parents" from the population and were used to "breed" two child chromosomes. This was carried out using crossover, which involves the swapping of sections of genetic material at random. The premise is that two well-performing parents may produce an even better-performing child. Further variation was introduced by a process known as mutation, in which small changes to the chromosome are made with random probability.

The operation of the GA continued over several generations during which time the less-well-performing chromosomes were purged from the population. Eventually, a steady state was reached in which no further improvement could be seen.

Inherent in the above process is the requirement for a performance metric to evaluate how well the filter resulting from each chromosome was performing. This property is known as "fitness." In this case, the fitness is a measure of the similarity of the filtered and clean images. The work here compares two different measures of fitness. The first measure used a weighted combination of the mean-absolute error (weighted at 0.6) and the mean-square error (weighted at 0.4). The second used a structure similarity (SSIM) index.[10] The SSIM is a metric developed for use in estimating the effect of subjective viewing of structural integrity. In all other respects, the details of the training process were identical.

Once the fitness of the different filters resulting from the child chromosomes had been measured, the best fifteen filters were kept and others discarded. The remaining chromosomes were then subjected to the GA techniques of crossover and mutation to create new chromosomes from which to generate new filters. These measures were used in an attempt to create new filters containing the desirable features of the successful filters and combine them to get closer to a near-optimal filter. The use of mutation introduces "new" information to the filter chromosomes. It allows areas of the search space to be reached that were not accessible using crossover alone.

The training runs were initially set to terminate after 30 minutes. After this time, the GA had completed 35 iterations. Figure 9.7 shows the improvement made on two examples of the training data by the best filter found using the MAE/MSE measured after 35 iterations. Clearly the SMF has led to an improvement, but there is still significant noise in one of the images. The GA was then set to run for 500 iterations.

The results from the filter produced after 500 iterations using the MAE/MSE hybrid as the error measure is shown in Fig. 9.8. The final error measure at the end of the training run was 0.996. The results of the filter created by using the SSIM as the fitness measure is shown after 35 iterations in Fig. 9.9 and 500 iterations in Fig. 9.10. The final error after 500 iterations was 0.991. Table 9.1 summarizes the quality differences between the images and the clean data by showing the improvement in the two quality measures after 35 and 500 iterations of the GA are applied to the training set.

It should be pointed out that the absolute values of the two measures are not directly comparable since the SSIM is much more closely related to the way in which

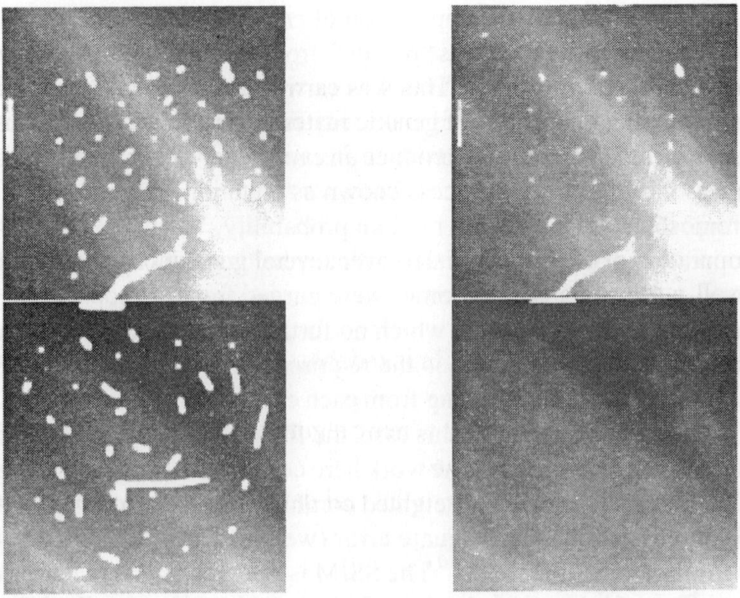

Figure 9.7 Output after 35 iterations of training using MAE/MSE hybrid measure.

Table 9.1. This table shows the improvement in the two quality measures after 35 and 500 iterations of the GA applied to the training set. The absolute values of the two measures are not directly comparable because the SSIM is much more closely related to the way in which the image is perceived by the human visual system. It can be seen, however, that SSIM makes a much greater improvement and continues to improve significantly between 35 and 500 iterations. On the other hand, the measure based on the mean-absolute and mean-square error makes very little further improvement after 35 iterations.

	Original image	After 35 iterations	After 500 iterations
MAE/MSE	0.9844	0.9957	0.9960
SSIM	0.7468	0.9242	0.9910

the image is perceived by the human visual system. However, it can be seen that SSIM makes a much greater improvement and continues to increase significantly between 35 and 500 iterations. Conversely, the measure based on the mean-absolute and mean-square error makes very little further improvement after 35 iterations.

The filter parameters resulting from the GAs after 500 iterations are shown in Figs. 9.11 and 9.12. As can be seen, these two filters have very different parameter values.

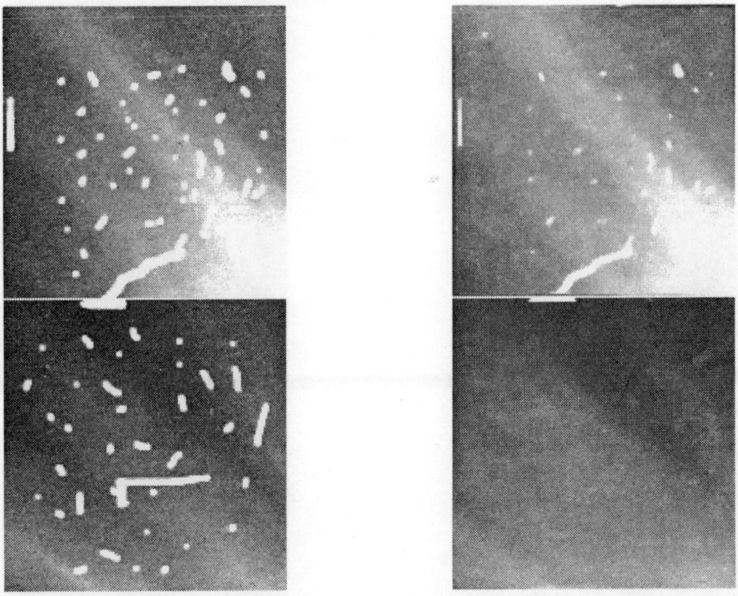

Figure 9.8 The filter output after 500 iterations. This filter was trained with a quality measure based on a weighted combination of the MAE and MSE.

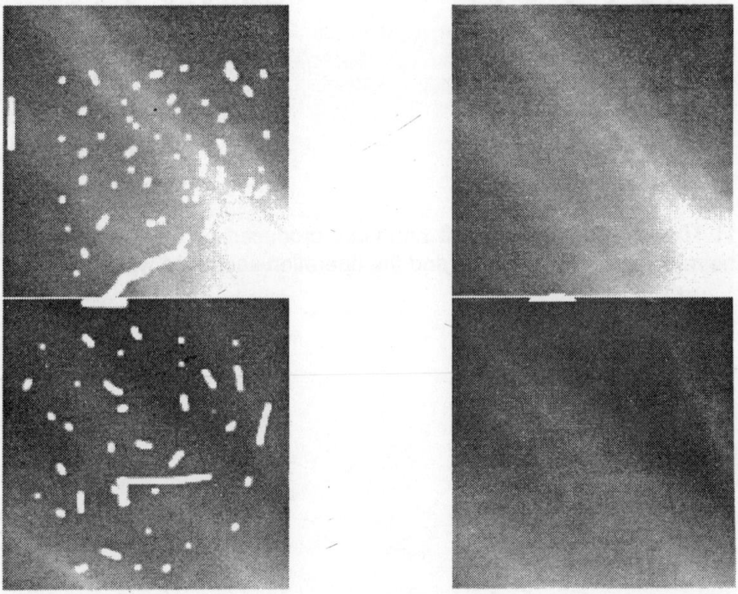

Figure 9.9 Filter output after 35 iterations trained using SSIM.

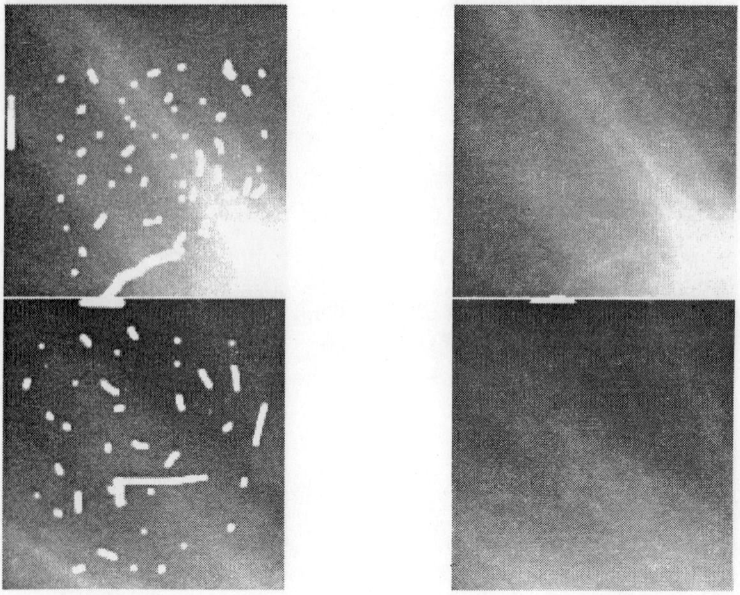

Figure 9.10 The output of the trained filter obtained after 500 iterations using the SSIM as a quality measure.

*	*	*	*	*
*	*	*	*	*
*	*	6	*	*
*	*	*	*	*
*	*	*	*	*

1	*	*	*	1
*	*	*	*	*
*	*	*	*	*
*	*	*	*	*
1	*	*	*	1

Figure 9.11 The hard center and soft boundary produced by the MAE/MSE training sequence. The rank for the filter was 17 and the operation sequence was a single erosion.

*	*	*	*	*
*	*	*	*	*
*	*	5	*	*
*	*	*	*	*
*	*	*	*	*

*	*	*	*	*
*	*	4	*	*
1	1	*	1	1
*	*	4	*	*
*	*	*	*	*

Figure 9.12 The hard center and soft boundary produced by the SSIM training model. The rank was determined to be 4 and the operation sequence was 2 erosions in sequence.

9.3.3 Application to real images

The near-optimal filters produced from the two training runs were then applied to the real noisy astronomical images shown in Figs. 9.1 and 9.2. The results using the filter trained in the first run of the application, with the MAE/MSE criterion, are shown in Fig. 9.13. The subjective improvement is very obvious. The image, which had been almost completely obscured by noise, has had a significant amount of the disturbance removed. The results of applying the filter created in the second training run and using the SSIM criterion are shown in Fig. 9.14. By comparison, the second run appears to have created a better filter and has removed almost all of the speckle.

Figure 9.13 The result of applying filters trained using the MAE/MSE combination quality measure.

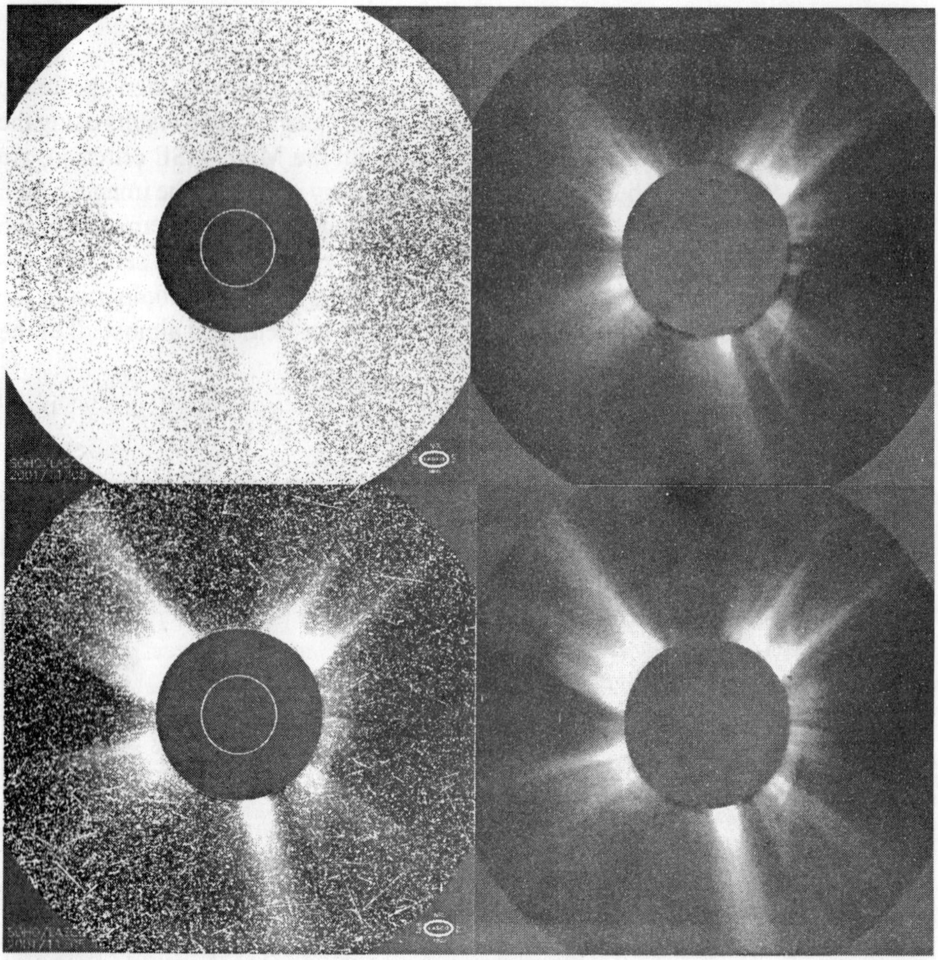

Figure 9.14 The result achieved by using the SSIM-based training algorithm. Minor artifacts can be seen, but otherwise the image is of a much higher subjective quality than the previous figure.

It is not possible to quote values of the quality measure for these images since they both require the original clean ideal images, which of course we do not have.

9.4 Hardware Implementation

The major processing step in the implementation of the soft morphological filter (SMF) is the rank ordering of data. For single images such as those shown in this chapter, the processing may be carried comfortably in software. The C++ program to process the 1024 × 1024 astronomical images in this chapter takes approximately 100 milliseconds per image to complete on a Pentium 4 running at 1.8 GHz with

512 Mb of RAM. There are two situations where hardware implementation may be of significant benefit: for the filter training process and for real-time implementation of spatio-temporal filters.

During training, the filtering process must be repeated many times and the quality criterion evaluated. In the example given, the first 35 iterations were completed in 30 minutes and 500 iterations took four hours and 20 minutes. Each iteration of the GA therefore took just under one minute to complete. This process could therefore benefit from hardware implementation. However, the training is usually carried out offline, so the longer processing times are not usually a problem.

A more challenging task is the implementation of a real-time spatio-temporal filter. This is because it must process a window that not only extends over 5 × 5 pixels, but also spans 3 frames. This requires that at least 75 pixels (depending on the repetition parameter) must be sorted in real time to find the *r*th largest or smallest value. If the processing involves multiple operations such as *soft erode* followed by *soft dilate*, then the intermediate values must be stored and processed again.

In order to test the viability of processing video streams in real time, two different strategies were implemented on a Xilinx-Virtex-II-based field-programmable gate array (FPGA).[11,12] The strategies differed only in the way they carried out the sorting process. The images were in approximate CIF format (360 × 280) and full 24-bit RGB color. In this case, the separate RBG color plans were filtered separately and recombined. This was found to give perceptually pleasing results, though this is contrary to popular wisdom. A comprehensive guide to the processing of color images is given in Sangwine.[13]

The two strategies implemented were called the partial-sort algorithm and the histogram algorithm.[14,15]

In the partial-sort algorithm, a traditional pairwise swapping approach[16] was used to obtain the maximum value in the set. This maximum value was then removed and the process repeated to find the 2nd largest value, and so on until the *r*th rank was obtained. For low ranks it was simpler to start at the bottom and work up.

The histogram algorithm, on the other hand, exploits the fact that the image data lies within a limited dynamic range. It is implemented by mapping the pixel values into a traditional histogram. The process of creating the histogram implicitly carries out a sorting operation in itself. The *r*th-ranked value is then simply determined by beginning at one end and counting the pixels in the histogram until the desired rank is obtained. The histogram approach is heavily dependent on the on-board memory used to accumulate the histogram whereas the partial-sort works purely with raw logic.

Both designs were synthesized using the Synplify 7.2 tool.[17] The target frequency used to set up the synthesis stage was 80 MHz. The partial-sort algorithm just managed to achieve this, whereas the histogram design was able to operate up to 110 MHz.

Figures 9.15 and 9.16 show the output of the MAP tool of the Xilinx ISE v6.1 software for the histogram and partial sorting algorithms, respectively. They show

```
Design Summary
--------------
Number of errors:      0
Number of warnings:   71
Logic Utilization:
  Total Number Slice Registers:          1,314 out of  28,672    4%
    Number used as Flip Flops:                    1,229
    Number used as Latches:                          85
  Number of 4 input LUTs:                1,409 out of  28,672    4%
Logic Distribution:
  Number of occupied Slices:             1,323 out of  14,336    9%
  Number of Slices containing only related logic:    1,323 out of    1,323  100%
  Number of Slices containing unrelated logic:           0 out of    1,323    0%
Total Number 4 input LUTs:               1,697 out of  28,672    5%
  Number used as logic:                  1,409
  Number used as a route-thru:             100
  Number used for Dual Port RAMs:          128
     (Two LUTs used per Dual Port RAM)
  Number used as Shift registers:           60

  Number of bonded IOBs:                   162 out of     720   22%
    IOB Flip Flops:                        224
  Number of Tbufs:                         256 out of   7,168    3%
  Number of Block RAMs:                     14 out of      96   14%
  Number of GCLKs:                           7 out of      16   43%
  Number of DCMs:                            4 out of      12   33%

Total equivalent gate count for design:  990,385
Additional JTAG gate count for IOBs:  7,776
Peak Memory Usage:  134 MB
```

Figure 9.15 Histogram resources utilization.

the resources taken up in the target FPGA and their utilization in the implemented designs.

The greatest difference between the resulting designs, in terms of resources, is in terms of the number of slices used. The partial-sort algorithm uses up to 87% of the whole FPGA while the histogram implementation barely uses 9%. Figure 9.17 shows the Floorplanner views of both designs. This is the allocation of resources on the device. It can be seen that the histogram uses far fewer resources.

These results are to be expected because the partial-sort algorithm only uses logic to perform the calculations. Consequently, it suffers from place and route problems. However, the histogram relies heavily on the use of memory blocks available on the FPGA.

This would not be the case if the designs were aimed at an ASIC. The equivalent gate count parameter in Figs. 9.15 and 9.16 indicates the approximate size of an ASIC to implement the design. For the partial-sort, only 278,412 gates are required, whereas the histogram requires 990,385. This huge difference is due to the large equivalent gate size for the memory blocks used in the local histogram approach.

These parameters show that the histogram implementation is better-suited for the FPGA but that the partial-sort algorithm would be expected to outperform the histogram design in a hypothetical ASIC implementation.

In terms of throughput, the partial-sort is more efficient, processing each frame in less than half the clock cycles. The FPGA performance tests show that the average processing time per frame is 0.0029 seconds for the partial-sort algorithm and 0.0070 seconds for the histogram.

```
Design Information
------------------
Command Line   : D:/Xilinx6/bin/nt/map.exe -intstyle ise -p xc2v3000-ff1152-6
-cm area -pr b -k 4 -c 100 -tx off -o soft_morph_3d_map.ncd soft_morph_3d.ngd
soft_morph_3d.pcf
Target Device  : x2v3000
Target Package : ff1152
Target Speed   : -6
Mapper Version : virtex2 -- $Revision: 1.16 $
Mapped Date    : Thu Dec 18 12:31:44 2003

Design Summary
--------------
Number of errors:      0
Number of warnings:   72
Logic Utilization:
  Total Number Slice Registers:      11,567 out of  28,672   40%
    Number used as Flip Flops:                 11,477
    Number used as Latches:                        90
  Number of 4 input LUTs:            17,088 out of  28,672   59%
Logic Distribution:
  Number of occupied Slices:         12,613 out of  14,336   87%
  Number of Slices containing only related logic:   12,613 out of  12,613  100%
  Number of Slices containing unrelated logic:           0 out of  12,613    0%
Total Number 4 input LUTs:           17,776 out of  28,672   61%
  Number used as logic:              17,088
  Number used as a route-thru:          528
  Number used for Dual Port RAMs:       128
    (Two LUTs used per Dual Port RAM)
  Number used as Shift registers:        32

  Number of bonded IOBs:                162 out of     720   22%
    IOB Flip Flops:                     329
  Number of Tbufs:                      416 out of   7,168    5%
  Number of GCLKs:                        7 out of      16   43%
  Number of DCMs:                         4 out of      12   33%

Total equivalent gate count for design:  278,412
Additional JTAG gate count for IOBs:  7,776
Peak Memory Usage:  308 MB
```

Figure 9.16 Partial-sort FPGA resource utilization.

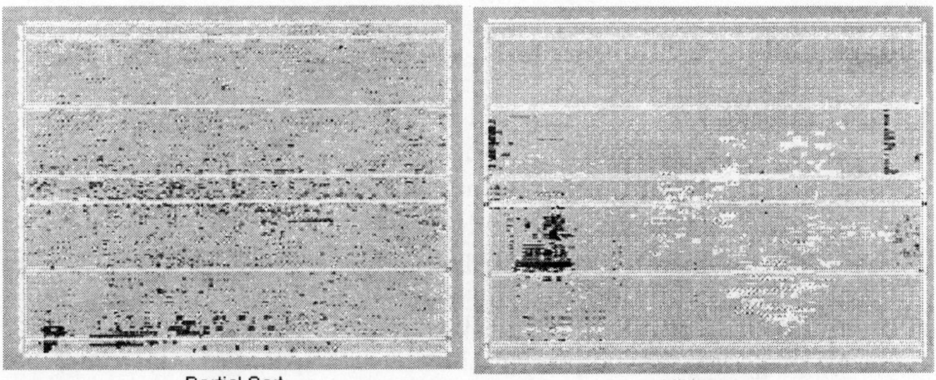

Partial Sort | Histogram

Figure 9.17 Floor plan of both designs. Light and dark parts on both images represent the area occupied by the filter block. Grey areas are unused.

Both designs were tested and the partial-sort reached a maximum frequency of 80Mhz, while the histogram was capable of working at frequencies over 100 MHz.

When the performance was measured overall (i.e., taking a video stream from the PC through the FPGA and back to the PC), the partial-sort algorithm achieved 13.1 frames per second (fps) while the histogram approach achieved 9.2 fps. The limiting factor was the overhead introduced by moving the data in and out of the FPGA device.

Another aspect to take into account is that the performance of the partial-sort design depends on the filter to be applied and the input stream, whereas the histogram performance is not affected by the filter nor the input stream and has the advantage of being adaptive to window sizes. The design is so small that it can easily be replicated on the same device to improve the performance.

9.5 Summary

This chapter has presented an overview of how the techniques introduced in the book may be used to solve a real-world problem. The use of a training algorithm is key to the creation of an appropriate nonlinear filter. In this case, the training set was produced by hand on a small image and this was used to obtain the optimal soft morphological filter. The filter itself was designed using a genetic algorithm run over 500 iterations. Two different quality criteria were compared. The resulting filters produced images with the noise significantly reduced and the structure intact.

The second half of the chapter considered approaches to hardware implementation of soft morphological filters in real time. Single images may be processed easily in software, so the presented example considered spatio-temporal images. It presented two approaches and gave performance metrics for each.

References

1 G. E. Brueckner, R. A. Howard, and M. J. Koomen, "The large angle spectroscopic coronagraph (LASCO)," *Solar Physics*, **162**, 357 (1995).

2 S. W. Kahler, "Solar flares and coronal mass ejections," *Annual Reviews of Astronomy and Astrophysics*, **30**, 113 (1992).

3 L. Koskinen and J. Astola, "Soft morphological filters: A robust morphological filtering method," *J. Electron. Imaging*, **3**, 60–70 (1994).

4 M. S. Hamid, S. Marshall, and N. Harvey, "GA optimisation of multidimensional gray-scale soft morphological filters with applications in archive film restoration," *IEEE Trans. Circuits and Systems for Video technology*, **13**(5), 406–416 (2003). See also **13**(7), 726 (2003).

5 P. Kraft, N. Harvey, and S. Marshall, "Parallel genetic algorithms in the optimization of morphological filters: a general design tool," *J. Electron. Imaging,* **6**(4), 504–516 (1997).

6 N. Harvey and S. Marshall, "GA Optimisation of Multidimensional Gray-Scale Soft Morphological Filters with Applications in Archive Film Restoration," *ISMM 2000* (2000).

7 S. Marshall, N. Harvey, and D. Greenhalgh, "Design of morphological filters using genetic algorithms," *EUSIPCO 2000*, Tampere, Finland (2000).

8 M. S. Hamid, S. Marshall, and N. Harvey, "GA optimisation of multidimensional gray-scale soft morphological filters with applications in archive film restoration," *IEEE Trans. Circuits and Systems for Video Technology*, **13**, 406–416 (2003).

9 J. H. Holland, *Adaptation in Natural and Artificial Systems*, MIT Press, Cambridge, MA (1995).

10 Z. Wang and A. C. Bovik, "A universal image quality index," *IEEE Signal Processing Letters*, **9**(3), 81–84 (2002).

11 Xilinx Inc., *Virtex2 Platform FPGA Handbook*.

12 B. Zeidman, *Designing with FPGAs & CPLDs*, CMP Books (2002).

13 S. J. Sangwine and R. E. N. Horne, *The Colour Image Processing Handbook*, Chapman and Hall, London (1998).

14 E. R Dougherty and J. Astola, *Introduction to Nonlinear Image Processing*, SPIE Press, Bellingham, WA (1994).

15 A. Gasteratos and I. Andreaidis, "Non-linear image processing in hardware," *Pattern Recognition*, **33**, 1013–1021 (2000).

16 R. Sedgewick, *Algorithms*, Addison-Wesley, New York (1988).

17 S. Esteban Zorita, *Implementation of Soft Morphological Filters Using FPGAs*, MPhil thesis, University of Strathclyde, UK (2006).

Chapter 10
Conclusions

This book has taken the reader on a journey through various image processing techniques, some of which will be new and some which will be familiar. On the way, we have encountered well-known methods such as the median filter, morphological operators and the hit-or-miss transform. Most other image processing texts start by deriving a filtering operator and mapping it to a finite sliding window. In this book we begin with the sliding window and consider the processing options available from it.

The values within the filter window are treated as logical inputs to a Boolean expression. The design process consists of identifying which Boolean expression (out of all those possible) will result in the lowest overall error. For binary images and small windows, the number of input combinations is sufficiently low such that the conditional probability of each output may be estimated accurately from a modest training set. The theory is straightforward and leads to simple methods for the calculation of the optimal filer and its associated error.

It is also easy to compare different filters and compute the increase in error for sub-optimal filters. The effects of the filters (in terms of which patterns of pixels are altered and which are left unchanged) can be seen to be consistent for additive and subtractive noise. For simple document-processing problems, the results can be stunning. This contrasts favorably with commonly used approaches of either applying the median filter regardless or heuristic filter design (i.e. guessing) at a pixel-processing level. The filter is defined in terms of an expression in Boolean logic that may be mapped directly to hardware.

The difficulty with this design approach comes when we wish to make the window size larger for more complex problems. For each extra location in the filter window, the number of input combinations doubles. Any window much larger than 3×3 pixels results in too many input combinations to estimate when using a training set containing a few images.

In any design approach based on training, it is essential that the size of the training set is matched to the complexity of the problem. If the search space is too large, constraints must be applied in order to limit the complexity of the problem. The ideal constraint is one that limits the search space so that a filter may be found with a finite training set, but allows sufficient flexibility to find an accurate solution to the problem. This is the point where intelligent human intervention is required in the process, rather than at the pixel level.

The error resulting from the design of a constrained filter has two components, constraint error and estimation error. The constraint error occurs as a result of limiting the filter complexity. The filter has fewer options available and so an increase in error may occur. The estimation error occurs as a result of the trained filter not having converged to its final value. For a fixed-sized training set, the estimation error gets smaller with a reduction in filter complexity.

The introduction of a constraint therefore increases the constraint error but reduces the estimation error. As with all engineering design problems, a trade off is involved in minimizing the overall error. In practice the estimation error can be very severe, even resulting in filtered images that are worse than the original. It is much easier to reduce the estimation error by adding a constraint (at the expense of an increased constraint error) than it is to do so by increasing the size of the training set.

Filter constraints can take many different forms. The earlier examples in this book assumed that the output for each input combination was estimated independently, which is reasonable for small windows. For larger windows independent estimation is almost impossible, therefore assumptions must be made about some inputs by considering others. This is equivalent to fitting a function in linear filtering. The simplest constraint on the function involves limiting the filter to increasing functions. Once a filter has been designed for a specific task, its performance can be evaluated a number of ways, such as by viewing the filtered output or analyzing the MAE figures. However, an interesting insight into the behavior of the resulting filter can be found by use of Boolean logic reduction techniques. The final optimized output function of the filter can be minimized into a sum-of-products form. This can be viewed as a set processing masks (consisting of black, white, and don't-care terms) to show how the patterns of pixels are changed by the processing.

Where the resulting function is an increasing filter, the processing masks correspond to morphological structuring elements. The sum-of-products expression is therefore equivalent to a union of erosions in morphology. In many applications of morphology, the structuring elements used are arrived at by heuristics, and it is unlikely that they are optimal in these circumstances. The statistical approach used in this book is ideal for producing optimum structuring elements for a given task.

Where the resulting function is nonincreasing, it is not sufficient to test if the structuring elements "fit" the foreground of the image. The hit-or-miss transform must be used to determine if corresponding conditions are met for the background. This would be the case for target recognition or OCR, for example. In any case, the optimal structuring elements are produced by this approach.

Several well-known filters may be expressed in the framework which may be used to design filters by this approach. Among them are weighted-order statistics filters including rank-order filters, the median, and its variants. Each of these filters involves a constraint that is achieved by restricting the possible logic functions that may be used.

The techniques described may be extended to grayscale images through a number of approaches. The issue of training and estimation error is further compounded by this extension to grayscale since each pixel would have at least 8 bits. The most straightforward way to extend these techniques to grayscale is by using threshold decomposition in a technique known as stack filtering. The input signal is split into a "stack" of binary signals, each of which may be processed by a binary filter. This binary filter may be estimated from a training set. The output from the binary filters is then restacked to produce a grayscale output signal. Many useful operators including rank-order, median, and some linear filters fall into the class of stack filters, as well as grayscale morphology with flat structuring elements.

More complex grayscale operations may be implemented in a framework known as computational morphology. This is inherently suited to digital implementation in a fixed number of bits. Its structure is similar to the stack filter because it produces a series of stacked binary outputs. However, the filtering operation is more complex and is based on a technique known as elemental erosion. Computational morphology is capable of implementing any operation, linear or nonlinear within the window size chosen. The result of designing a filter in this framework is a kernel of structuring elements. In the most general case, these are unrelated, other than the fact that they must observe an ordering to avoid violating the stacking property. For grayscale morphology and stack filters, the kernels are related in a simple way.

The design of filters based on computational morphology is difficult because they are so general in nature and their search space is very large, requiring unrealistic amounts of training data.

A simplification of computational morphology is known as the aperture filter. This is based on a windowing process in the amplitude domain of the signal, as well as the spatial domain. Signal points lying outside of the window are simply clipped. As a result of their reduced dynamic range, aperture filters may be designed with training sets of realistic size. They have been successfully used in many applications including deblurring and object recognition. Aperture filters have been extended to multiscale applications. The most difficult problem remains that of aperture placement.

A useful technique that reduces gross errors in nonlinear filtering is called envelope filtering. In this case, the output of a filter is forced to be contained between and upper and lower limits of a bounding envelope. This has the advantage that the error is never larger than the envelope's range.

Designing image processing operators in terms of digital logic results in solutions that can be transferred straight to hardware without further mapping or truncation. Frequently the implementations resulting directly from the design methods described here are cumbersome. For example, in the case of stack filters it is expensive to duplicate the processing hardware for each of the 256 levels present in an 8-bit grayscale image. Fortunately, other techniques may be used to reduce the hardware required. Examples of the implementation of stack filters, computational morphology filters, and aperture filters have been presented.

The penultimate chapter covers the specific case of noise removal from astronomical images, showing how the use of a training algorithm is essential to the creation of an appropriate nonlinear filter. With a manually-produced training set, it was shown how an optimal soft morphological filter could be obtained. The chapter also gives an example of real-time FPSA implementation of a spatio-temporal filter.

The objective of this book was one of translation: from the language of mathematics and set theory to the language of electronics and computer science. Many of the powerful techniques outlined in this book are not yet in common use for industrial image processing. The author believes that this is because these techniques have been developed primarily by mathematicians, and their descriptions reside in texts that are not easily accessible to those who build industrial image processing kits.

Ideally this book has gone some way to correcting that situation. Some complex areas have necessarily been glossed over, but if you wish to know more about the complexities, there are plenty of good texts. A good starting point would be Dougherty and Barrera's paper (referenced below) which bridges the gap between pattern recognition theory and nonlinear signal processing.[1] I hope you have enjoyed our journey, and thanks for reading to the very end.

Reference

1 E. R. Dougherty and J. Barrera, "Pattern recognition theory in nonlinear signal processing," *Mathematical Imaging and Vision*, **16**(3), 181–197 (2002).

Index

S

T

U

W

X

Professor Stephen Marshall was born in Sunderland, England. He received a first- class honors degree in Electrical and Electronic Engineering from the University of Nottingham in 1979, and a PhD in Image Processing from University of Strathclyde in 1989. Between his degrees he worked at Plessey Office Systems, Nottingham, University of Paisley, and at the University of Rhode Island.

Prof. Marshall is Head of the Image Processing Group in the Department of Electronic and Electrical Engineering at the University of Strathclyde. In recent years, his research activities have been focused in the area of nonlinear image processing. During this time, he has pioneered new design techniques for morphological filters based on a class of iterative search techniques known as genetic algorithms. The resulting filters have been applied as 4D operators to successfully restore old film archive material. This work is now in the process of commercialization. Recently he has applied these techniques in the area of genomic signal processing.

He has published over 100 conference and journal papers on these topics in many publications, including those of IEE, IEEE, SPIE, SIAM, ICASSP, VIE and EUSIPCO. He is currently an Associate Editor of the European Signal Processing Society's (EURASIP) *Journal of Applied Signal Processing* (*JASP*).

In 2000 he was admitted as a Fellow of IEE and is a founding member of their Professional Network (PN) in Visual Information Engineering (VIE). He is currently chairman of the Technical Advisory Panel of the VIE PN.

Prof. Marshall is a Law Society of Scotland approved Expert Witness. His advice has been sought in many court cases in which video evidence has been critical. He is also a former Director and Chairman of the Scottish Chapter of the British Machine Vision Association and also a former member of the IEE Professional Group E4 in Vision, Image, and Signal Processing.